CHANGES IN GLOBAL CLIMATE

A Study of the Effect of Radiation and Other Factors
during the Present Century

CHANGES IN GLOBAL CLIMATE

A Study of the Effect of Radiation and
Other Factors during the Present Century

K. Ya. Kondrat'ev

RUSSIAN TRANSLATIONS SERIES

26

1992
A.A. BALKEMA/ROTTERDAM

Translation of:

Radiatsionnye Faktory Sovremennykh Izmenenii Global'nogo Klimata.
Gidrometeoizdat Publishers, Leningrad, 1980

© *1985 Copyright reserved*

Translator: Gyan Arora
Editor: V. Pandit

ISBN 90 6191 442 6

UDC 551.5

Possible variations in the solar constant, gaseous composition and aerosol content of the atmosphere are among the most important external factors contributing to contemporary changes in global climate. These factors have been discussed in detail in the present monograph. The author has also examined the properties of atmospheric aerosol and its possible influence on climate; such an influence is manifest in the form of variation in the earth's albedo and influx of radiant heat through the atmosphere. The author has also studied the effect of anthropogenic factors (primarily of halocarbons and their derivatives) on the ozone layer and their influence on the influx of radiant heat in the stratosphere.

Contents

Introduction

3   The study of climate and of its variations has become one of the most
important disciplines in modern science. The importance of the subject
and of related problems has increased considerably in view of the significant changes in the surrounding environment as a result of human
activity and its probable impact on the climate. This becomes more
important as climatic variations (for instance, prolonged drought in the
Sahelian region of Africa and its disastrous consequences) seriously affect
agricultural production and various aspects of human industrial activity.

In recent years, investigations concerning the possible impact of
supersonic aviation, space transport vehicles, halogen-containing hydrocarbons and other factors have invoked considerable interest. The significant stability of aerosol pollution of the stratosphere has stimulated a
study of the possibilities of influencing the aerosol layer of the stratosphere
in order to prevent undesirable climatic changes. The impact of nuclear
tests on weather and climate and problems of heat releases are also
discussed. All this attributes exceptional importance to the study of the
physical factors of climate and its variation and, thereby, to the further
development of the theory of climate.

It is well known that the main difficulty in developing the theory of
climate as well as in revealing the most important factors which cause a
climatic change lies in the existence of several such factors and their
complicated mutual interactions. For this reason, discussions are continuing relating to the definition of the concept of climate. Discussions on
the predictability of climate are also of considerable importance.

Climate is determined by the statistical characteristics of an ensemble
of different levels of the atmosphere and the underlying surface. That is
why, when an actual atmosphere is replaced by an idealized mathematical
system, it is assumed that the ensemble involves all possible levels over
an infinite interval of time. In such a case, climate by definition is invariable and the concept of its predictability and prediction becomes un-
important. The interest in climatic changes is however determined by the
4 fact that the statistical characteristics of the atmosphere and the under-

1

lying surface during a sufficiently long interval of time can differ considerably from the characteristics of a subsequent similar interval. Therefore, it is expedient to define the concept of climate in terms of an ensemble of levels for a continuous, though finite interval of time.

A clear understanding of the reasons for contemporary climatic changes depends, to a considerable degree, on data from the biological sciences in regard to the conditions of vegetation and soil and geographical investigations. An analysis of the consequences of an increase in carbon-dioxide concentration and atmospheric dust-loading due to industrial pollution is of great importance. For a meteorologist, it necessitates some specialization in the field of the natural environment. Further development of the theory and realization of numerical models describing the circulation of a number of components of the natural environment (carbon, nitrogen, sulfur) which may affect the climate, are important.

The wide use of not only the usual methods but also of space observations becomes necessary to monitor the parameters, important from the viewpoint of the theory of climate. Satellites can be used, particularly for investigation of pollution of the atmosphere and ocean and observing such changes on the land surface occurring as a result of human activity, which may have an effect on the climate.

Important stages in the progress of investigations of climate and assessment of future perspectives were the World Conference on Climate followed by the Fifth World Meteorological Congress. The World Conference on Climate held in Geneva from February 12–23, 1979, adopted a declaration on the problems of global climate. The declaration asserts that though mankind is exploiting the advantages of a suitable climate at present, the industrial activities of man have been subjected to the effect of a variability of climate, especially to extreme phenomenon such as droughts and floods. In the first instance, these relate to the developing countries, located mostly in the arid and semiarid zones or regions of excess humidity.

The interrelation of climate in different countries determines the necessity of a coordinated global strategy in the interest of better understanding and rational exploitation of climate. At present man is introduc-
5 ing unforeseen changes in climate at local and to a certain extent at regional levels. As further expansion of industrial activity may lead to considerable regional and even global climatic changes, there is therefore a serious concern over the need for international cooperation.

Changes in climate under the influence of natural processes will continue. The trend toward a gradual cooling in some parts of the northern hemisphere in the last several decades is similar to natural changes of the past and it is not yet known whether this will continue. Though significant progress in the understanding of climatic variations has been achieved, there still exists considerable uncertainty regarding the roles of various factors. Probably the continuing anthropogenically caused increase of carbon dioxide causes a warming of the lower atmosphere, especially at higher latitudes. It is also possible that some effects of this warming on a regional or global scale will be observed even toward the end of this century, and toward the middle of the next century these effects would become significant.

Examples of manifestation of such anthropogenic influences, are various aspects of agricultural practices (increasing application of nitrogen fertilizers), forestry, release of halocarbon compounds, etc. From the viewpoint of processes on dry land, the consideration of changes in the surface roughness and albedo, thermal characteristics of the surface soil, and the moisture-retention capacity of the soil have been found to be of great importance.

An assessment of the consequences of population growth and man's industrial activities points to the possibility of unexpected climatic changes on the planet. These changes may become comparable with major natural variations taking place during a span of several human generations. Estimates of present-day tendencies of climate indicate that they would be determined predominantly by natural factors at least till the end of the current century. An analysis of the localized climatic consequences of war or of possible large-scale climatic changes as a result of a worldwide thermonuclear conflict is extremely important.

In view of the fact that investigation of the interaction between the atmosphere and ocean is of great importance in the study of climate, preference should be given to the study of those regions of the ocean, specified under the program 'Cross sections', where the interaction is highest (regions of emergence of the Gulf Stream and the Kuroshio Current, zones of unstable stratification near the edges of ice cover, upwelling
6 regions and monsoonal zones). Besides the aforementioned meteorological, hydrologic and geophysical information, corresponding data on agricultural practice, land management, industrial production as well as ecological data is essential.

Thus, acquisition of a sufficiently comprehensive array of data on the climatic system (atmosphere-ocean-land surface-cryosphere-biosphere) covering a period of not less than 30 years, should become the main objective. Tracking of anthropogenic factors and their effects require special attention. All these effects should be assessed by expanding the system of visual and satellite observations. In this connection detailed investigations on dry land with consideration of albedo, evaporation, evapotranspiration of the earth's moisture, roughness of the land surface, snow and ice cover, which are associated with the problems of anthropogenic influences on climate, assume significant importance. The question of the effect of aerosol on the radiation regime of the atmosphere and cloud formation is also important. The development of a set of models of atmospheric aerosol, which can be used in numerical modeling of climate, plays a key role in the given conditions. No doubt, a leading role is played by the effect of cloudiness on climate and this requires formulation of corresponding methods of parameterization taking into account all significant feedback links. However, investigations on predictability and sensitivity of climate should form the basis of our study.

If information is to be extracted from long-period internal fluctuations of the atmosphere-ocean-land surface-cryosphere system, then, among numerous possible factors determining contemporary climatic changes, the three main factors are:

1) variation of solar constant, 2) transformation of properties of the underlying surface, and 3) changes in the gaseous and aerosol composition of the atmosphere.

Hence, the question regarding the effect of variation in composition of the atmosphere on climate appears to be at the center of the problem relating to the interpretation of contemporary climatic changes. The physical content of this question is reduced to the problem of the effect of variation in the composition of the atmosphere on radiant heat influx, in which the issues related to the study of cloudiness, aerosol and optically active gases in the atmosphere assume an important place. After a brief description of observational data on contemporary climatic changes and discussion on the problems of the earth's radiation budget, we should concentrate mainly on a consideration of the results of investigations on the atmospheric aerosols and their effects on radiative transfer and the possible role of the stratosphere in climatic changes considering that, for a number of reasons, these questions have become especially important.

Chapter I

Contemporary Global Climatic Changes and Radiation Budget of the Earth

7   Before entering into a discussion on the radiation factors of contemporary global climatic changes, let us consider the basic parameters characterizing observed changes in the climate. In these discussions we would define climatic changes during the last hundred years as 'contemporary'.

1. TRENDS OF CONTEMPORARY CLIMATIC CHANGES ACCORDING TO OBSERVATIONAL DATA

1.1. Air temperature. Different types of data have been responsible for the present understanding (which is in fact no more than a conjecture), regarding the evolution of climate during the long history of the earth. Meteorological observations during the last hundred years enable us to experimentally establish the existence of climatic changes. For example, there has been an increase in the mean global temperature by 0.6 °C during the years 1880 to 1940, followed by a trend toward a decrease in temperature [7, 11–13, 15, 18, 37, 38, 50–53, 58, 59, 63, 72, 73, 78, 97, 121–126, 182]. According to the 'WMO Statement on Climatic Changes' [197], there exists a probability of further cooling, although the cumulative effect of anthropogenic factors should cause warming. Borisenkov and Priemov [5], and Agee [58] also arrived at a similar conclusion. Investigation of reasons for such a tendency (particularly, an explanation of the fact as to whether this is a consequence of industrialization) is of tremendous importance for mankind, since in the case of continuation of presently observed cooling an ice age may set in over a period of 200 to 300 years.

On the other hand, assumptions regarding the exponential growth in

5

industrially produced heat and considerations of an increasing concentration of carbon dioxide in the atmosphere lead to the conclusion of a possible catastrophical warming of the climate. Budyko and Vinnikov [6] have formulated exactly such a forecast in their work. Based upon the tendency, observed from the mid sixties, of a warming of the climate 8 (Fig. I.1, see also [117, 121]), these authors have built up a forecast of significant global warming by the year 2000. In view of the inconsistency of conclusions regarding possible climatic changes in the future and even in regard to climatic trends during the last decade, we will discuss the latest results of analysis of observational data, published during recent years.

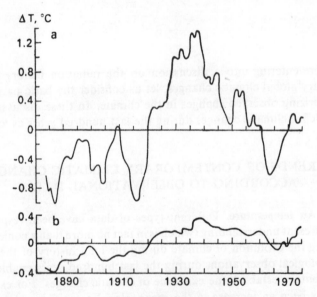

Fig. I.1. Secular course of air temperature anomalies in northern hemisphere
(5-year running average):
a) 72.5–87.5°N zone; and b) 17.5–87.5°N zone.

Angell and Korshover [63] analyzed variations in the mean annual temperature at the level of the earth's surface and also in the 300–250 and 300–100 mb layers on the basis of radiosonde observations from 63 stations. The stations have been selected as representative of seven climatic zones (northern and southern polar regions, temperate and subtropical latitudes of both hemispheres and the equatorial belt) and are suitable from the standpoint of continuing the observations in future. In [63] it has been shown that during the period from 1958 to 1965 cooling prevailed over the major portion of the earth, lowering temperature by about

0.3 °C, but starting from 1971 onward, a slight warming was observed at almost all latitudes. It may be noted that till 1965 there had been a continuous increase in the meridional temperature gradient between tropical and temperate latitudes, subsequent to that an opposite tendency emerged with a warming near the surface which was particularly significant in the Antarctic.

9   In the northern hemisphere an increase in the vertical temperature gradient, especially above Asia, was observed, whereas in the southern hemisphere the variation of the temperature gradient appeared to be erratic. In the tropical troposphere, temperature fluctuations with an amplitude of 0.3 °C over a period of about 3 years prevailed after 1965, which started damping toward 1975. A possible consequence of the eruption of Mount Agung in 1963 was a decrease in surface temperature by about 0.2 °C in tropical and southern extratropical latitudes and a cooling almost twice as great in northern extratropical latitudes. Apart from this, in the year of eruption, in temperate latitudes of the southern hemisphere, there was a warming in the 300–100 mb layer and a cooling in the 300–250 mb layer by about 1 °C.

The variability of temperature at the surface and in the 300–250 mb layer has increased by 15% during the period under consideration, from station to station, indicating conditions of enhanced 'extreme' weather. In the zone including Australia and New Zealand a more intense warming and an insignificant cooling were observed.

Harley and Jacobsson [97] investigated climatic trends in various regions of the northern hemisphere during a period of 26 years (1949–1974) on the basis of data on the thickness of the 1000–500 mb layer, averaged over an area of 5° of latitude and 10° of longitude from 25°N to the North Pole (for the Pacific Ocean only, the data from 1964 onward has been used). An analysis of a series of maps of five-year averages of thickness revealed the localization and extent of regions of stable cooling and warming.

Four regions of cooling have been identified: a small region in the north and a large one in the eastern part of Asia, a small region in America (Wisconsin), a small region in the central part of the Atlantic and only one region of warming in Iran. Most severe cooling took place in Mongolia (Ulan-Bator region), and maximum warming in Mauritania (after 1970).

The consideration of five-year running averages led to the identification of three major temperature fluctuations with maximums in 1958, 1962–63 and 1967 respectively; these corresponded (approximately) to maximums of the running averages for the winters (1958, 1962 and 1966–67). An approximation of seasonal trends over a 22-year period, by quadratic polynomials led to the conclusion of the existence of continuous

8

negative and positive trends, observed mainly during winter in eastern
Asia and northern Iran respectively. The maximum amplitude of
cooling in the northern hemisphere prevailed in the eastern and northern
parts of Asia, particularly during summer and spring.

10   Centers of warming and cooling can be distinctly seen on the seasonal
maps of mean temperature trends over the territory of Canada. These
specific features are related with the cold summer seasons and severe
freezing in the Canadian Arctic, as well as with periods of abnormally
low surface temperatures and atmospheric pressure over the United
States of America.

The map of mean annual temperature variations for the northern
hemisphere specifically indicates the existence of distinct spatial structures of zones having different tendencies. The peculiarities of their
geographical variability are caused by the distribution of continents and
oceans, effects of orography and ice cover and also by periodic variations
of atmospheric circulation.

Analysis of mean annual trends and monthly values of the thickness
of the 1000–500 mb layer in the 30–80°N latitudinal belt over the period
1949–74 leads to the conclusion of a decrease in average temperature of
the lower troposphere by 0.3 °C during the said period. In the opinion of
the authors of [97], such a comparatively small change cannot serve as a
sufficient basis for forecasting a new ice age in the foreseeable future.
Detailed examination of the aforementioned trends led to the observation
that in the background of a general tendency of cooling, a severe warming was observed in all latitudes at the end of 1950 with a maximum in
1950, another maximum in 1959 (year of maximum solar activity), and a
sharp minimum in 1965. A strong correlation exists between the decrease
in average thickness of the 1000–500 mb layer and the decrease of surface
temperature in the northern hemisphere since 1938.

Damon and Kunen [78] carried out an analysis of the climatic trend
of air temperature near the earth's surface on the basis of data from 67
meteorological stations located in the southern hemisphere for the period
1943 to 1974. They averaged the observational data over five-year intervals, corresponding to the middle parts of the last three 11-year solar
activity cycles. When the authors used data of only those stations for
which observations were available continuously during the period 1943
to 1974, and excluded stations for which either the locations were changed
or observations were available for not more than 10% of the annual data,
there remained only 57 stations. The data from these stations does not
show any appreciable hemispherical trend of temperature for the period
considered (it is important to note that 53 of these 57 stations are located
in the 0–45°S belt) but it does indicate trends of warming or cooling of
individual regions. For example, cooling is observed in subequatorial

Africa, but warming in Australia and New Zealand. In the cities with po-
11 pulations of more than 750 thousand an increase in temperature by 0.2 °C
was observed due to the urban 'thermal island' effect. Observations from
15 stations located in the zone of 45–90°S revealed a warming by 0.12 °C
from the five-year period 1955–1959 to the five-year period 1960–64, and a
subsequent increase in temperature by 0.37 °C toward the next five-year
period 1970–74. The process of warming is more pronounced at higher
latitudes. However, in at least one high-latitude region (South Orkney and
the Falkland islands) cooling occurred.

According to the authors of [78], the data considered encourages us
to be cautious about the generally held view of global cooling from 1940
onward. It seems that this cooling is limited only to the northern hemisphere and some regions of the southern hemisphere. The cooling at
higher latitudes of the northern hemisphere, to a known extent, is balanced by a warming at higher latitudes of the southern hemisphere starting
from 1955–1959.

There can be two possible explanations for the observed climatic
trends: 1) the trend is due only to natural factors and remains practically
independent of anthropogenic influences and 2) the trend is caused by a
combined effect of anthropogenically increased carbon-dioxide concentration and enhanced solar activity after 1964–1969, which resulted in a
warming in the southern hemisphere. Possibly this effect does not appear
in the northern hemisphere due to the severe cooling caused by aerosols
of anthropogenic origin.

The complexity of the phenomena determining weather and climate
and lack of observational data do not permit us at present, to precisely
judge whether or not anthropogenic factors exert any considerable effect
on the climate. Therefore, immediate efforts are required for long-term
global monitoring of climatic parameters and their determining factors.
In this case the observational data at higher latitudes (more than 60°) is
particularly significant.

The results of processing of observational data on secular variations
of average temperature of the 1000–500 mb layer in the northern hemisphere on the basis of data by Boville [71] (including data on recent years)
are presented in Fig. I.2. This data shows that after 1975 there was a
considerable fall and subsequent increase in the average monthly thickness of the 1000–500 mb layer toward the end of 1977. The year-to-year
variability in climate observed over the last 10–15 years, undoubtedly
indicates the presence of a trend toward cooling over the last three decades (Fig. I.2, curve 2). The exceptionally important fact of increased
variability of climate is more explicitly expressed (curve 3).
12   Kukla et al. [138a] carried out an analysis of secular trends of various
climatic characteristics on the observational data for the last 25–30 years.

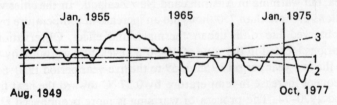

Fig. I.2. Secular course of thickness of 1000–500 mb layer for 25–90°N
latitudinal belt: 1) mean monthly thickness of 1000–500 mb layer;
2) secular trend of layer thickness (mean temperature); and
3) secular trend of variability of layer thickness (mean
temperature).

The characteristics under study include air temperature near the earth's
surface, mean temperature of the 1000–500 and 1000–100 mb layers,
temperature of the ocean surface, extent of the zone of the circumpolar
vortex at the 300 mb level in the northern hemisphere, and the extent of
snow and ice cover, taken together as well as individually. The results of
analysis for the northern hemisphere are most reliable, as in this case a
sufficiently representative array of data is available. The conclusions
relating to the southern hemisphere and the entire globe should be considered as purely tentative.

The results, thus obtained primarily indicate the existence of considerable inter-year variability of all climatic characteristics. This variability
forms a dominant part of the total variability of the averaged zonal
values under consideration. Inter-year variability is in phase for most
parts of the northern hemisphere and many regions of the southern
hemisphere. In a majority of cases the amplitudes of inter-year fluctuations of climatic characteristics exceed by many times their secular trends,
thereby creating serious difficulties in detecting these trends. A reliable
identification of change in the sign of trends is possible only by the use
of data from parallel observations of various climatic characteristics over
fairly long intervals of time.

The data for lower and temperate latitudes of the northern hemisphere reveals a cooling accompanied by fluctuations during the complete
period under consideration. Cooling at higher latitudes continued till
1965 followed by a temperature stabilization and warming. Clear
evidence of global cooling was seen in the years 1954–1956, 1964–1965,
1968, 1971–72, 1974 and 1976. On the other hand, the winters of 1958,
1959, 1970, 1973 and 1975 were relatively warmer. During the last ten
13 years several temperature minimums occurred in the northern hemisphere and the 0–50°N belt; these touched the level of the lowest minimums observed in the second half of the fifties.

Minimums of ocean surface temperatures in the seventies were significantly lower than in the fifties. During the period 1950–1975 the rate of cooling in the northern hemisphere, characterized by various climatic parameters, varied within the limits of (0.1–0.2) °C/10 years. The steepest falls in the ocean surface temperature and in the temperature of free atmosphere were recorded in the northern parts of temperate latitudes and at higher latitudes of the Atlantic Ocean. The decrease in air temperature in the vicinity of the ocean surface was relatively small. Due to wide fluctuations in temperature this trend cannot be considered statistically significant.

The magnitude of changes in five-year running averages of the majority of climatic characteristics (variations during five years) for temperate and lower latitudes of the northern hemisphere has increased in 1971–75 as compared to the years 1966–70. However, at higher latitudes a decrease and even a change in the sign of trends has been observed. The period from 1971 to 1975 was characterized by a fall in surface temperature of the northern belt of the Pacific and the Atlantic oceans, a considerable increase in the stretch of snow cover and in the zone of circumpolar vortex and a significant drop in the average temperature of the free atmosphere at lower and temperate latitudes as compared to the previous five years. Record anomalies occurred in 1976 and 1977.

The data under consideration does not reveal a transition from cooling to warming in the northern hemisphere. The short-term warming of 1973 and 1975 was not repeated in 1976. The air temperature near the surface in spring (March-May) during the last ten years has increased at higher latitudes of the northern hemisphere, but noticeable trends at lower and temperate latitudes have not been observed. At the same time a gradual expansion of the zone of circumpolar vortex and snow cover took place. For the autumn temperatures of the last decade a significant decrease in higher latitudes, with a slight drop in temperate and lower latitudes, is typical.

The extent of autumn snow cover increased considerably, but the changes in the circumpolar vortex in autumn and winter were comparatively slight. The data for the southern hemisphere, though quantitatively insufficient (especially for the 30–60 °C belt), shows a cooling till the mid sixties, which was later replaced by a slight warming. In higher latitudes south of 60 °S) the air in the free atmosphere gradually became warm during the entire period under consideration. The area of pack ice increased during 1967–1973, but later started decreasing.

The dynamics of climate in both hemispheres reveal a certain parallelism: warming of the atmosphere at high latitudes during the last ten years; increase in the extent of snow and ice cover in 1970 and 1972, but a decrease in 1974 and 1975; and acceleration in the cooling trend at the beginning of the sixties. In other words, it may be considered that a

12

variable cooling prevailed during the last 30 years in the northern hemisphere. A change in the direction of trends (transition of warming) was not observed. Hence, further investigations of space-time nonuniformity of climate are of immense importance.

Concluding the brief discussion on observational data relating to secular variations of temperature, the following two conclusions may be drawn:

i) till the present day, the trend toward cooling of the global climate continues (particularly in the northern hemisphere); and

ii) an increase in variability of the global thermal regime is evident.

Though investigations on contemporary global climatic changes undertaken during recent years mainly concentrate on an analysis of temperature trends, it is obvious that the variability of other climatic characteristics is no less important. These are: the humidity of air and soil, precipitation, distribution of snow and ice cover, radiation budget and its components, etc. The importance of analysis of observational data concerns such climatic factors as ozone content and concentrations of carbon dioxide, aerosol and other optically active minor components of the atmosphere. Taking into account the limited information available regarding most factors, we shall limit further discussion to a brief preliminary review of the most important and interesting information on the extent of snow and ice cover and the concentration of carbon dioxide in the atmosphere.

1.2. Snow and ice cover. The considerable success in obtaining information on the global snow and ice cover can be attributed to progress in the field of satellite meteorology, particularly to the applications of microwave remote sensing methods [2–4, 8–10, 14, 19, 21, 22, 28, 32, 34, 36, 41, 48, 49, 55, 65, 74, 93, 94, 116–118, 132–135, 137, 156, 158, 166–168, 189, 190].

Since the time of the first investigations of snow cover and the ice situation, on the basis of analysis of televised maps of the earth obtained from the meteorological satellite *Tiros-2*, in the spring of 1961, the scope of such satellite observations has significantly increased owing to
15 the improved spatial resolution of images and use of multispectral images for wavelengths of from 0.4 μm to 1.55 cm. Thus, for example, infrared images of the Antarctic region on a wavelength of 3.8 μm, obtained in 1964 from the *Nimbus-1* satellite enabled us to trace gaps in the ice cover (as temperature contrasts) during periods of polar night. Data from an infrared radiometer of higher resolution in the 0.7–1.3 μm band positioned on *Nimbus-3* proved to be useful for identification of snow and ice melting zones. A series of pictures from satellites, such as *Meteor*, *Nimbus* and *NOAA* opened ways for tracking the dynamics of snow cover and the use of this data for forecasting river flow.

Gloersen and Salomonson discussed in detail the results in the field of satellite glaciology, recently obtained from analysis of the *Landsat-1* satellite data on natural resources and microwave images on the 1.55 cm wavelength from the *Nimbus-5* satellite. The interpretation of data from *Landsat-1* concerning the Wind River basin (State of Wyoming, USA) showed that by using the empirical relationship between the extent of snow cover and discharge, forecasting of a flow is possible from satellite data (empirical relationships appear to be different, depending upon the height of the catchment area above sea level). Images from the *Landsat-1* satellite; with a spatial resolution of about 80 m, proved to be especially useful for the study of the structure and dynamics of glaciers. In [93] examples of application of brightness spectra for identification of individual elements of glaciers are illustrated.

From the results of observation, obtained with the help of instrumentation installed in *Nimbus-5*, the images obtained with the help of a triple-channel (8.3–9.3; 10.2–11.2; 0.8–1.1 μm) radiometer for mapping properties of natural formations (RMPNF), and microwave images on the 1.55 cm wavelength, are most informative from the point of view of glaciologic applications. The images recorded by the RMPNF have a resolution of about 660 m and can serve as a valuable aid for tracking snow and ice cover, if clouds do not intervene.

Microwave images are the sources of information on year-round snow and ice conditions (only in the presence of a zone of intensive precipitation is the basic contribution to the outgoing radiation in the 1.55 cm wavelength, determined by the atmosphere). Simultaneous use of microwave and infrared (10.5–12.5 μm) measurements opened up possibilities 16 for drawing maps of the emissive power of ice cover. To be more precise, reference is made here to the ratio of brightness temperatures, one of which, namely the radiobrightness temperature, determines the properties of the top soil having a thickness of the order of a few wavelengths.

An analysis of the radiobrightness temperature maps for Greenland and the Antarctic for the first two years of operation of *Nimbus-5* indicated the reliability of their use for characterizing the ice cover (including the age of ice), but did not enable the detection of any noticeable changes in emissive power from year to year. In Greenland, severe seasonal changes related with melting and freezing, have been revealed. In this connection a successful experiment on mapping of the boundaries of melting of snow was taken up.

The microwave scanning radiometer, operating on a 1.55 cm wavelength, installed on the meteorological satellites *Nimbus-5* and *Nimbus-6* enabled us to obtain data necessary for plotting the maps of polar ice (the spatial resolution of the images is 32 km). Gloersen et al. [94, 137] analyzed the radiothermal images of the northern and southern polar caps of the earth relating to the winter of 1972–73. For the available measurement accuracies, the ice-cover compaction, in the given case, is determined with an error not exceeding 6%.

The analysis of images of the polar zones (three images from each series) enables us to reveal the following: 1) the existence of considerable divergence between the data of various climatic atlases on ice cover of the polar regions and the actually observed distribution of ice cover, 2) a noticeable difference in distribution of perennial ice in the northern polar zone with the distribution predicted on the basis of existing models of ice-cover dynamics, 3) irregularities of the perennial ice edge in the Antarctic, which were neither observed before nor were ever predicted, and 4) unique contours of isolines of radiobrightness temperature of Greenland and Antarctic glaciers, which are determined, apparently by the morphological structure of the snow and ice cover; moreover in the Antarctic the variability of the physical temperature of the upper layer of ice appears to be an important factor.

The dynamics of polar ice testifies to the necessity of tracking from satellites the changing (and not corresponding to the climatic data) ice situation. Very low radiobrightness temperatures have been recorded in the regions of Greenland (155–200 K) and the Antarctic (130–160 K). In the Antarctic, an appreciable correlation between radiobrightness temperatures and temperatures of the upper layer of the continental ice cover is observed. In Greenland however, such a correlation is absent.

17   In order to develop a reliable method of remote sensing of ice-cover properties, aircraft measurements using very sophisticated instrumentation and particularly complex satellite measurements, which directly determine the properties of ice, have great significance. The Joint Soviet-American Bering Sea Experiment (BESEX) is an example of implementation of such a program [55].

Microwave images can be effectively used for the study of ice cover morphology and dynamics round the clock and in any weather conditions. Satellite microwave images are especially important for investigation of variations in polar ice cover. As has already been mentioned, they opened up the possibility of revealing serious divergences from the data of atlases of polar ice. An analysis of images, obtained from side-looking radars in aircraft, created possibilities of their interpretation not only for the purpose of revealing the comprehensive morphology and dynamics of ice, but also for the estimation of ice cover thickness as well as a more detailed differentiation of types of ice.

The measurement data from the 22.2 and 31.4 GHz channels of a microwave spectrometer (MS) installed on the *Nimbus-5* satellite were used to obtain information on global distribution and certain properties of snow and ice cover. The spatial resolution of the microwave spectrometer at the level of the earth's surface covers about 200×300 km².

As shown by Künzi et al. [118, 166], for the interpretation of data under study, it is significant that the process of internal scattering considerably affects the radiothermal emission of the majority of snow and ice types. In [166] the mean radiobrightness temperature $\overline{T}_b = \frac{(T_{b;\,22.2} + T_{b;\,31.4})}{2}$ K, and gradient $\varDelta T_b = \frac{(T_{b;\,31.4} - T_{b;\,22.2})}{2}$ K/GHz, have been accepted as the parameters characterizing the distribution of snow and ice cover [166]. The measurement accuracy of absolute values of radiobrightness temperature is about 2 K, and for relative values, it is better than 1 K. Therefore, the mean square errors in the absolute values of $\overline{T}_b$ and $\varDelta T_b$ are approximately equal to 2 K and 0.2 K/GHz, and in relative values equal to 1 K and 0.1 K/Ghz.

Under typical conditions of humidity of the atmosphere and emittance of snow and ice, the corrections to $\overline{T}_b$ and $\varDelta T_b$ due to the effect of the intermediate atmospheric layer are less than 4 K and between 0.1 and -0.1 K/GHz respectively. Since these corrections are relatively small as compared to the variations of studied parameters, they were not taken into account.

The observational data for the winter and summer of 1973 in both polar regions is reproduced in [166]. An analysis of this data establishes 18 the distinct characteristic features of the mean radiobrightness temperature and its gradient in the case of snow, and oceanic ice, as well as continental ice in Greenland and the Antarctic, when the extent of snow and ice cover is sufficiently wide (of the order of 10^5 km² or more). The seasonal variations of snow and ice cover are clearly observed. The use of data from double-channel measurements enable identification of various types of oceanic ice and firn. For mapping of snow cover on land and oceanic ice, multifrequency remote sensing can be used.

Further modification of the methods of interpretation of data requires a theoretical modeling of microwave properties of snow and ice. More specifically, this is required from the point of view of developing more comprehensive models taking into account scattering and data concerning the subsurface temperature profile. The ground controlled measurements as well as laboratory investigations of microwave properties of snow and ice can be of immense importance in improving the reliability of interpretation of the data.

An important parameter, characterizing the snow cover, is its water equivalent. This knowledge is of significant importance for forecasting the flow. The earlier results show that the water equivalent correlates with the emissivity of snow cover in the microwave band.

In [118] maps of snow cover on the basis of data for several five-day periods from December 1972 to December 1973 have been drawn, but only for wide zones in Europe, Asia and North America due to the low spatial resolution of the microwave spectrometer (about 170 km). The snow-cover maps drawn with the help of computers reveal that the extent of snow cover in December 1972 was much higher than in December 1973. This can be illustrated by the data (Table I.1) on the portion of European territory falling under snow cover (30–60°E, 50–63°N).

Table I.1. Portion of European territory falling under snow cover

Year	Snow cover (days)	Portion of territory, %
1972	Dec. 23–28	67
1973	Jan. 21–27	59
	Apr. 15–18	5
	Jun. 17–21	0
	Aug. 22–26	0
	Dec. 22–26	45

The limited data available does not yet provide a means for evaluating the water equivalent of snow cover. In this connection ground investigations of specific features of microwave radiation of snow cover in the frequency band of 1–100 GHz are being planned. The interpretation of observational data with the help of a scanning microwave radiometer, installed on the *Nimbus-6* satellite, is of great interest.

The scanning microwave radiometer (SMR), mounted on the *Nimbus-6* satellite is basically meant for continuous global mapping of the underlying surface on frequencies of 22.235 and 31.400 GHz at a spatial resolution of 150 km. Stepwise scanning with the help of SMR is accomplished for six angles (on both sides of the satellite trajectory) within limits of ±53° scanning angles with respect to the nadir.

Fisher et al. [158] reproduced global microwave images illustrating the possibilities of tracking the seasonal evolution of polar caps of the earth. The areas of perennial sea ice are distinctly differentiated by their comparatively lower emissivity and its decrease with frequency ($\Delta T = T_{b;\ 31.4} - T_{b;\ 22.2} > 0$). As a rule, in the case of sea ice $\Delta T < 0$, and only in winter are small positive values observed. The radiobrightness temperatures in the Antarctic are considerably lower than surface temperatures, since the emissivity here varies in the range 0.6–0.9.

The structure of the radiobrightness temperature field almost corresponds to the profiles of snow accumulation. The zone of weak accumulation in the center of Greenland is clearly distinguished. In the Antarctic, yearly variations with positive values in summer decrease and a transition to small negative values in winter is characteristic for ΔT. From the field of ΔT, it is possible to identify firn in the background of ice. Land covered with ice is characterized by lower emittance and negative values of ΔT.

Microwave radiometers with electrical scanning (MRES) set up in the *Nimbus-5* and *Nimbus-6* satellites enabled us to obtain images of the underlying surface on the 1.55 and 0.81 cm wavelengths, respectively, with a spatial resolution of about 30 km. This opens up, particularly, the possibility of tracking peculiarities of spatial distribution and dynamics of Antarctic sea ice caused by ocean currents. In order to test the GARP observation system, Zwally and Gloersen [201] undertook drawing weekly maps of sea ice, characterizing the distribution of fractions of open water (%) in terms of rectangles of $2.5 \times 2.5°$ latitudinal and longitudinal segments.

Individual ice fields and pools of open water usually have dimensions smaller than the resolution of MRES. However, determination of the ratio between ice cover and open water within the limits of each segment of resolution seems possible in view of the significant difference in the emittances of ice and water. The observed radiobrightness temperature T_b can be written in the form of a linear combination of the brightness temperature of ice (εT_0) and water ($\varepsilon_w T_w$) taking into account the contribution of atmospheric radiation (A):

$$T_b = (\varepsilon T_0) C + (\varepsilon_w T_w + A)(1 - C),$$

where C is the ice concentration. Typical values of various parameters on the 1.55 cm wavelength are:

$$\varepsilon_w T_w = 120 \text{ K}, \quad A = 15 \text{ K and } \varepsilon T_0 = 250 \text{ K}.$$

The determination of ice concentration from the measurement data on one frequency becomes complicated due to the variability of ε in relation to the type of ice, the absence of independent measurements of ice surface temperature T_0 and variability in microwave radiation of the atmosphere. Since the ice in the Antarctic zone is, as a rule, annual and changes in T_0 are relatively small, determination of the ice-cover concentration (compaction) with an accuracy of $\pm 15\%$ or more is possible.

The location of the southern boundary of sea ice in the Antarctic zone during winter is regulated primarily by the circumpolar currents and not by predominant winds. This boundary clearly reflects the effect of underwater ridges on the current. The existence of two stable zones of lower compaction of winter pack ice confirms the existence of upwellings of warm water ($T \approx 2$ °C) toward the surface. The ice concentration in the Ross Sea is lowered, for example, by 15–30%. But an upwelling is clearly seen as a large extended air hole covering an area of 0.25×10^6 km^2 near the Greenwich meridian, which was firmly conserved during the winter of 1974 and also that of 1975. Ice concentration here was not less than

15%. Since the heat and moisture exchanges over an air hole are several orders higher in value than those over pack ice, the presence of such an air hole should have a significant impact on the regional climate.

High albedo, attaining values of up to 0.80. of snow and ice, and the large amount of snow and ice on the globe (in the regions of polar latitudes) make the extent of snow and ice cover one of the most important factors in climatic formation. From paleoclimatological data, it is understood that the ice-to-water ratio in the Arctic has undergone considerable changes during the last thousand years. If during the period 860–1200 A.D., the average period of annual occurrence of drifting ice near the Greenlandic shores was zero and rarely exceeded 1–2 weeks, then in the subsequent hundred years it varied in the limits of 0–5 weeks and during the period 1600–1900 A.D. rose to 15–25 weeks. It again decreased almost to zero by the year 1950. The decrease in the area covered under ice during the periods of warming from 1900 to 1940 A.D. was about 3×10^6 km².

21   Recent investigations relating to subpolar oceans have revealed that in the period 1840 to 1900 there were one or more 'discharges' of shelf ice (mainly in the Weddel Sea) at least of the order of 10^5 km², which caused a significant cooling (about 5 K on Illusion Island). Apparently, the extent of ice cover increased up to 2×10^6 km².

On the basis of the model of radiative-convective equilibrium, developed by Manabe and Wetherald, Flöhn [57] calculated changes in the equilibrium temperature of both the hemispheres and the entire globe under the influence of the aforementioned variations in the extent of ice cover. Decrease in the equilibrium temperature of the northern and southern hemispheres at the maximum extent of ice cover during the last hundred years was 0.95 and 0.60 K respectively. The temperature decrease of 5–6 K, as obtained from the 'model' for the last glaciation (25000–15000 B.C.) has been found to correspond with the available paleoclimatological data.

It may be assumed, that the real variations of mean temperature in the past have closely followed the variability of the global extent of ice cover. (This does not, however, mean that the changes of albedo under the influence of variations in the extent of ice cover should be considered as a reason for large dimensional fluctuations of the climate). Leaving aside various hypotheses regarding the genesis of climatic variations, it may be concluded that surprisingly large variations in the extents of ice cover in the polar regions of both hemispheres are of significant importance in understanding and forecasting climatic variations. Hence, the importance of mapping the extent of ice cover in both the hemispheres from satellites is evident.

The data obtained by G.J. Kukla and H.J. Kukla [116, 117] demonstrates that in the early seventies the snow cover and pack ice first appeared in the northern hemisphere and were more widespread than seven years later when regular mapping of snow and ice cover by satellite began. The said changes appeared to be more significant in 1971 and undoubtedly must have been reflected in the thermal budget of the hemisphere. The anomalous weather conditions in 1972, and 1973 may have been a consequence of such changes. These conclusions on the dynamics of snow and ice cover were derived as a result of the analysis of weekly maps compiled by the National Oceanic and Atmospheric Administration on the basis of images of the earth from meteorological satellites [117]. In these maps snow and ice fields preserved for at least a day, have been recorded.

22 An analysis of data for the period from March 30, 1967 to November 1, 1973, shows that a minimal area of snow and ice cover was observed in the northern hemisphere till 1971. After this, it was observed from the end of August till the beginning of September in 1973 [$(9.5–9.8) \times 10^6$ km^2] and in 1972 (10.3×10^6 km^2). In the summer of 1973, the Arctic Ocean was more ice-free than during the previous six years. The most significant contrast (increase) of the mean monthly area, covered under snow and ice in the northern hemisphere occurred during February-April, and September-November in 1970 and 1971 respectively. This increase during October and November continued till 1972. In October, 1972, the area of snow and ice was twice as much as in 1968. The determination of running averages of yearly values indicated that during 1971 this area increased by 12% (from 32.0×10^6 to 36.9×10^6 km^2) and since then fluctuated within the limits of $(36.7–37.5) \times 10^6$ km^2. standing at 36.7×10^6 km^2 on November 1, 1973. Hence, the year 1971 may be considered as a turning point from the standpoint of snow- and ice-cover regimes.

In the work of G.J. Kukla and H.J. Kukla [117] the possible sources of errors in the obtained results have been discussed and it is indicated that these errors do not affect the conclusions on the variability of snow and ice cover from year to year. For evaluation of the effect of the snow and ice cover dynamics on the reflection of shortwave radiation by the underlying surface an arbitrary index of reflection $R = fMQS$ has been introduced, where $f = 0.30$ is the coefficient characterizing the differences in the albedo of the underlying surface in the case of absence and presence of snow and ice $M = 0.67$ the coefficient which takes into account the attenuation of radiation by the clouds; Q the total amount of radiation during the day for a cloudless sky, and S the area of snow and ice cover.

The data for seven years in the northern hemisphere indicates that the index R is maximum during spring and minimum during autumn. During the spring of 1969 and 1973 the maximum values of this coefficient were observed. As in the case of snow and ice cover, in 1971 a rapid increase of R took place. On November 1, 1973, the index R was equal to 8.6×10^{19} J/day. It is known that, during the period from 1968 till 1971 the basic characteristics of atmospheric circulation were close to normal, whereas during the summer of 1972, and the winter and spring of 1973 the weather was abnormal in many respects. In this connection, a correlation between the increase in the area under snow and ice cover and decrease in the average temperature of the 500–1000 mb layer in 1971–1972 has been observed. This is more clearly expressed in the lower latitudes (during the same period when the air temperature above the North Pole increases).

However, further investigations would be necessary for the confirmation of the aforementioned correlation. The mean annual area of the snow and ice cover in the northern hemisphere during the period of the last 23 severe winter is estimated at $(60–70) \times 10^6$ km^2, which is more than the present value, by approximately, 30×10^6 km^2. Since in 1971 the area of snow and ice increased by 4×10^6 km^2, seven such years would be enough for attaining the conditions of an ice age. Hence, there exist possibilities of rapid climatic changes, and tracking of the dynamics of the global snow and ice cover appears, therefore, to be an exceptionally important task.

Since the beginning of the functioning of the meteorological satellites in 1960, the regular (weekly) tracking of the location of the southern boundary of the snow cover became possible. The first maps characterizing the dynamics of this boundary in the northern hemisphere, were drawn in 1966 on the basis of data from the *ESSA-3* satellite. Shortly afterward the weekly maps of snow- and ice-cover dynamics found an application in functional weather forecasting. Wiesnet and Matson [189, 190] compiled, on the basis of data for nine years (1966–1975), averaged over a month (December-March), maps of snow and ice cover distribution in the northern hemisphere. They also obtained the average (over several years) maps for the months under consideration. The location of the snow- and ice-cover boundaries have been brought out on maps at a scale of 1 : 50000000 in the polar stereographic projections by the authors. For this the images obtained in the visible and also in the infrared regions of the spectrum served as sources of information. By way of planimetering for every individual month, they determined the area of the snow cover for North America and Eurasia. For latitudes south of 52°, the low level of solar illumination or its total absence during the winter in the latitudinal belt of 60–90°N hinders our obtaining reliable data. In this case the accuracy of estimation of the area of snow and ice cover is 5–10%.

An analysis of variation of the ice cover area led to the results, which practically coincide with the results obtained by G.J. Kukla and H.J. Kukla [117]. The variability of snow cover in Eurasia contrasts with the relative stability of snow cover in North America. Therefore, as a rule, a correlation between ice cover in Eurasia and North America does not exist. Only in December and March is an approximate phase coincidence in the variations of snow covered area observed.

During the entire nine-year period under consideration, no noticeable changes in the snow covered area were observed in North America. The maximum extent of snow cover (the data relates to North America in the region south of 52°N) constituted 6.4×10^6 km^2 in January 1970 and the minimum, 2.3×10^6 km^2 in March 1968. Though a noticeable inter-year variability is observed, a secular trend is absent. December and March are especially characterized by a considerable variability in snow cover. The wide extent of snow cover in Eurasia determines its dominating effect on snow cover variability in the northern hemisphere.

On the one hand, during three winters from 1968 till 1971 an increase in the area under snow cover took place and on the other, an opposite trend was observed during three winters from 1972 till 1975. In Eurasia, the maximum extent of snow cover was observed in February 1968 (15.4×10^6 km^2) and in February 1972 (15.7×10^6 km^2), and the minimum (6.9×10^6 km^2) in February 1970.

The entire nine-year period is characterized by cyclic variations of the snow covered area. This confirmed the fact that there is no justification in using data obtained over short periods for forecasting long-term trends. Apparently, the albedo feedback, which is given significant importance in a number of works relating to the theory of climate, can have a pronounced effect on the short-term variability of snow cover. The mapping of snow cover dynamics from satellite data is of great interest and it would be worthwhile to compile a summary of observational data for every five years for both the hemispheres as well as for the Arctic and the Antarctic during summer.

The average values of the extent of continental snow cover for the period of observation from 1966 till 1976 are presented in Table I.2. This data should be considered at present as the most representative of continental snow cover climatology for Eurasia and North America.

The analysis of data on the evolution of ice cover in the Arctic, carried out by Volkov and Zakharov [9], in view of the climatic changes, led to the conclusion regarding a significant increase in the ice cover, starting from the forties of the current century, which has continued even in recent years. The authors [9] anticipate, a further intensification of cooling and increase in the area of ice cover in the Arctic Ocean approximately till the nineties.

Summing up, it should be emphasized that presently the application

Table I.2. **Average extent of mean monthly snow cover over Eurasia and North America for 9 or 10 years**

Month	Period of averaging (No. of years)	Mean value, $\times 10^6$ km^2	Mean square deviation, $\times 10^6$ km^2
		Eurasia	
Dec	9	20.0	1.9
Jan	9	23.0	2.0
Feb	10	23.7	3.0
Mar	10	19.8	2.3
		North America	
Dec	9	13.7	1.0
Jan	9	14.8	0.5
Feb	10	14.4	0.5
Mar	10	13.1	1.0

of multispectral images of the underlying surface in the visible, infrared and microwave regions of the spectrum make possible a regular mapping of snow and ice cover, and permit us to study (depending upon the spatial resolution of images) the mechanism of its detailed spatial structure. No doubt, further developments in microwave remote sensing with the use of direct or indirect methods will lead to the creation of a foundation for year-round mapping of spatial distribution and state of snow and ice.

In order to develop quantitative methods of interpretation of data, comprehensive aircraft and satellite investigations are of great importance in corroborating ground-based measurements, and the corresponding laboratory and theoretical methods.

1.3. Carbon dioxide. This gaseous component of the atmosphere attracts maximum attention because of the problem in regard to its anthropogenic impact on the climate [33, 64, 67, 109, 143–145]. Thus the study of factual data regarding the trend of carbon-dioxide concentration is of interest. Since carbon dioxide is always considered as an external factor affecting the climate, it is important to clarify how far such an assumption is valid.

At present, there exist most convincing proofs concerning anthropogenic impact on the natural global cycle of carbon dioxide. Since the time of the beginning of the industrial revolution an almost continuous increase in the anthropogenic impact of carbon dioxide in the atmosphere has been observed, as about 90% of energy generation is related to the use of fossil fuels (coal, gas, petroleum). During the last 30 years, the yearly increase in the carbon-dioxide content constituted 4.8%. During 1959-71, the carbon-dioxide concentration has increased from 313 to 323 ppm. The results of direct measurements confirm that half of the carbon dioxide

26 remains in the atmosphere and the other half goes into the World Ocean, the biosphere and to a certain extent into sedimentary rocks.

An extrapolation of existing trends and analysis of the geochemical cycle of carbon dioxide leads to the conclusion that during the next 80 years, a fourfold increase of carbon-dioxide content in the atmosphere should occur, if the presently observed trend continues. Analysis of oceanic sedimentary rocks in order to evaluate the variations of accumulation of carbonates of calcium can serve as a key to understanding the natural changes of the carbon cycle in the course of an ice age and especially over the last 20000 years.

Calculations show [109], that about 18000 years ago the alkalinity of the ocean constituted 3.06 meqv/1, the carbon-dioxide content amounted to 2.39 mmol/ 1 and the pH was 8.6. These conditions correspond to a partial pressure of carbon dioxide in the atmosphere equal to 163 ppm, which is about 50% of the present-day value. Though the results of the calculations under consideration are extremely arbitrary, they serve as an example illustrating the possibilities of obtaining similar more reliable estimates in future on the basis of more accurate information on the accumulation of carbonates in the Pacific and Indian oceans.

The possible impact of an increasing carbon-dioxide concentration in the atmosphere on climatic changes in the future requires more attention toward prognostic estimates of an increase in the carbon-dioxide content. Till recently, an increase in concentration has been attributed, as a rule, to the burning of fossil fuels. However, in recent years it has become more and more evident that such an activity as felling of forests is of no less importance. In this connection arises the important problem of the study of the global carbon cycle in the context of the biosphere.

It is known that 99% of the global biomass (826×10^9 t of carbon) is accounted for by the continents and that is why it may be subjected to burning [153]. Thus, it is important to note that 93% of the biomass is constituted by forest ecosystems and only 7% is contributed by savanna, grass cover, tundra deserts, agricultural forests and other ecosystems. The accumulation of dead organic matter (detritus) which amounts to 1460×10^9 t (including 55×10^9 t of surface detritus, readily available for burning) is closely related with the living biomass. According to another estimate, the global content of detritus is 3000×10^9 t, but possibly this is somewhat underestimated.

The total quantity of 2286×10^9 t of carbon characterizes its global reservoir in the earth's (continental) ecosystems. Since this considerably
27 exceeds the atmospheric repository (670×10^9 t), it is natural that changes in the earth's reserve may have a considerable impact on the atmospheric reserve. One of the most significant sources of variability of the earth's reservoir of carbon is fires, intensification of which results in an increase

in the carbon-dioxide content of the atmosphere.

A decrease in biomass as a result of fires is more significant when forest ecosystems, characterized by a high potential biomass, are transformed more or less into grass or shrub ecosystems with a smaller biomass. If, for example, the usual biomass range of deciduous forests in temperate latitudes is 3–100 kg of carbon per square meter, then for grass cover, a biomass range of 0.1–2 kg of carbon per square meter would be typical.

Quantitative assessment of changes in the frequency of forest fires over the last several thousand years is not possible. Reiner and Wright [153] analyzed the probable effect of a change in forest-fire frequency and other factors on the variations of the biomass and carbon repository in North America and for the globe as a whole. This analysis led to the conclusion that during the last several thousand years the amount of carbon in the earth's ecosystem has undoubtedly decreased. The main reason for this was, apparently, deforestation and subsequent burning, and not an increase in the frequency of forest fires.

The greatest changes in the ecosystem and carbon reservoir are apparently related to the destruction, mainly in the nineteenth century, of primary forests in the area of which cultivable land and secondary forests with smaller biomass appeared. The global drop in the biomass, relative to its present-day value, constitutes about 20% (165×10^9 t), which is a considerable fraction of the atmospheric carbon reserve.

It should be accepted that the trend toward a biomass decrease is continuing even today, but corresponding quantitative assessments are highly contradictory: $(0.5 - 3.6) \times 10^9$ t of carbon per year. The figures, thus obtained call for serious attention to the worldwide concern over an increasing use of wood as a fuel (from global figures for 1972 it appears that the fraction of wood used as fuel stands at 65%). The consumption of wood as fuel, felling of forests, intense mineralization of soil, and soil erosion should be considered as major factors which determine the global variations of carbon reserves.

A thorough analysis of the role of the biosphere in the global carbon dioxide budget was carried out by Woodwell [198]. The measurement of concentration of the ^{13}C isotope in tree rings serves as an important indicator of long-term changes in carbon-dioxide content. According to calculations of industrial releases of carbon dioxide into the atmosphere, the cumulative release is equivalent to 60×10^9 t of carbon from 1850 to 1950. The total carbon-dioxide content in the global atmosphere is less accurately known. In 1850 it was $(540–625) \times 10^9$ t, which corresponds to a relative carbon-dioxide concentration of 260–300 ppm [198].

A specific feature of carbon dioxide of anthropogenic (industrial) origin is the absence of the ^{14}C isotope in it (this isotope is formed in the composition of natural carbon dioxide as a result of an impact of neutrons generated by cosmic rays on atmospheric nitrogen, and after the year 1950 additional quantities of ^{14}C carbon emerged as a result of nuclear tests) and a decrease in the content of the stable ^{13}C isotope (this is related to enrichment with the ^{12}C isotope as compared to the ^{13}C isotope in the process of photosynthesis in plants). Since anthropogenic inputs from the burning of fossil fuels and biospheric production of carbon dioxide lead to a decrease of the ^{13}C concentration in the atmosphere and a decrease in the concentration of ^{14}C takes place only under the influence of anthropogenic sources, a comparison of ^{13}C and ^{14}C isotope dynamics enables us to assess the contribution of the biosphere to the changes in carbon dioxide content in the atmosphere.

An analysis of cellulose from tree rings confirms a gradual decrease in ^{13}C and ^{14}C concentrations during the period 1850 to 1950. The comparison of ^{13}C dynamics with that of ^{14}C led to the conclusion that a rise in carbon-dioxide concentration caused by the biosphere mainly applies to the period 1860-1930. Approximate estimates showed that the carbon dioxide release during this period, caused by biospheric changes, is equal to the cumulative anthropogenic releases during 1850–1970 and constitutes 120×10^9 t of carbon. The total amount of carbon dioxide entering the atmosphere during a hundred years from 1850 till 1950 is equivalent to 180×10^9 t of carbon.

The decrease in the global biomass which occurred during the hundred years under consideration constituted 7%. During the past several decades the rate of decrease in biomass has been further reduced. This has caused a severe decrease in the contribution of the biosphere to the release of carbon dioxide into the atmosphere. If during a hundred years from 1850 to 1950 two-thirds of the increase in carbon-dioxide concentration of the atmosphere were related to the biosphere, presently anthropogenic releases make a dominant contribution at the cost of burning fossil fuel.

The study of the migration of ^{14}C and ^3H isotopes, these being products of nuclear tests and considered to be indicators of the atomized carbon dioxide penetrate into the ocean as an important reserve of carbon dioxide, revealed the existence of two major zones of carbon dioxide releases into the upper displaced layer of the ocean, situated in the regions near 30°S and 30°N. About 34% of the carbon dioxide released into the atmosphere is dissolved by the displaced layer, and 13% by deeper layers of the ocean. Thus, about 53% of the carbon dioxide remains in the atmosphere. The total increase of carbon-dioxide concentration in the atmosphere from 1850 till today is about 18%. Since these conclusions heavily depend on the accuracy of data on the course of ^{13}C concentration (even a change of the order of 10^{-4} generates a variation of about 0.3×10^9 t of carbon), for a more reliable estimate of the contribution of the atmosphere in the changes of carbon-dioxide content more prolonged and accurate observations combined with model calculations are necessary.

The clearly expressed annual trend of carbon-dioxide concentration determined mainly by the maximum intensity of photosynthesis during summer in forests in temperate latitudes of the northern hemisphere, can serve as an indicator of the important contribution of the biota in the carbon-dioxide dynamics (Fig. I.3). No doubt, forests significantly affect the short-period variability of carbon-dioxide concentration, but it should be taken into account that felling of forests and oxidation of humus are also important factors in long-term patterns of increase in carbon-dioxide concentration.

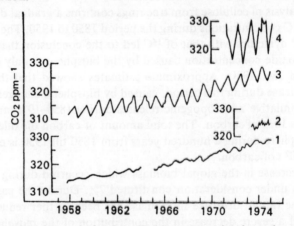

Fig. I.3. Secular trend of carbon-dioxide concentration from observational data at various points:
1—the South Pole; 2—Samoa; 3—Mauna Loa; and 4—Point Barrow (Alaska).

According to the data obtained by Woodwell [198] the global atmosphere presently contains 7.00×10^{17} g of carbon in the form of carbon dioxide which exists in a state of continuous exchange with the biota and upper layers of the ocean. The fraction of the biota is about 8.00×10^{17} g of carbon and, furthermore, a potential source of carbon $(1-3) \times 10^{18}$ g is the organic matter in the soil (mainly humus). Cutting of trees, extensive cultivation in the zone of soils containing large quantities of organic matter, and distribution of swamps have accelerated the process of humus disintegration accompanied by the release of carbon dioxide in the atmosphere.

The most important potential reservoir of carbon is the World Ocean with carbon dioxide (about 40000×10^{15} g) dissolved in it.

If we take into account the deep layers of the ocean, then it may be considered that over periods of thousands of years the carbon-dioxide content in the atmosphere is determined by equilibrium with inorganic carbon in the deep layers of ocean waters $(35000–38000) \times 10^{15}$ g, not taking into account the already formed sedimentary rocks. The upper displaced layer of the ocean with a thickness of about 100 m undergoes a more intense exchange of carbon dioxide with the atmosphere and hence, contains 600×10^{15} g of inorganic carbon.

Calculations of the global carbon cycle are rather contradictory [163–164]. One of the possible schemes of the carbon cycle proposed by Woodwell [198] is illustrated in Fig. I.4. The estimate of carbon dioxide released as a result of burning of fuels of the order of 5×10^{15} g per year and the corresponding increase of carbon-dioxide content in the atmosphere equivalent to 2.3×10^{15} g of carbon per year can be considered as fairly reliable. This means that certain processes should maintain a withdrawal

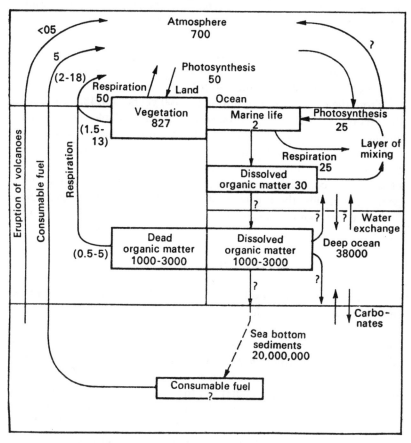

Fig. I.4. Global carbon cycle (in units of 10^{15} g). Second and third branches of cycle from the left demonstrate the effect of anthropogenic factors.

of 2.7×10^{15} g of carbon per year. The available estimates show that the adsorption of carbon dioxide by the ocean, apparently, cannot be so intense and can hardly exceed 2.5×10^{15} g per year.

Recent calculations led to the conclusions, that the main fraction of the products of photosynthesis does not belong to the oceans, but to dry land, and the major reservoirs of carbon are the tropical forests, playing an extremely important role in the global carbon budget. Cutting of forests (and even their replacement by artificial afforestation) can therefore appear to be a significant factor for an increase in the carbon-dioxide concentration in the atmosphere.

Recent estimations enabled us to conclude that the global biota is a source, and not a sink, of atmospheric carbon dioxide (Fig. I.4). According to approximate estimations an additional $(2–18) \times 10^{15}$ g of carbon in the form of carbon dioxide is injected into the atmosphere as a result of felling of forests and oxidization of humus. Thus the total equivalent delivery comprises $(7–23) \times 10^{15}$ g per year, out of which only 2.3×10^{15} g is accumulated in the atmosphere. Hence, it becomes necessary to explain the additional accumulation of large quantities of carbon dioxide (this problem still remains unresolved).

From the data of Stuiver, during the 100 years from 1850 to 1950, on the average, 1.2×10^{15} g of carbon was being released into the atmosphere annually to the detriment of the biota, whereas industrial deliveries of carbon dioxide are equivalent to 5×10^{15} g per year. The aforementioned data, despite its contradiction convincingly demonstrates the importance of the biotic contribution as a source of carbon dioxide in the atmosphere. From this, the urgent need for satellite tracking of the state of forests, undertaking measures for the conservation of forests and a further study of the problem of a global carbon cycle, is felt.

2. THE SOLAR CONSTANT

Determination of the solar constant (total flux of solar radiation just outside the atmosphere, derived on the basis of the average distance between the earth and the sun) has a fundamental importance not only for meteorology; during recent years precision measurements of extra-atmospheric spectral distribution of solar radiation became essential, for example, in connection with the necessity to reliably predict the thermal regimes of space devices. Therefore, the long history of investigations of the solar constant is understandable, a description of which can be found, for instance, in monographs [81, 111], review papers [29, 69, 90–92, 177–179, 200] and in a number of other publications. Fairly comprehensive reviews of data on extraterrestrial spectral distribution of solar radiation are contained in [39, 68, 79, 98–103, 119, 138, 165, 191].

Several years of investigations, carried out toward the end of the nineteenth and beginning of the twentieth centuries in the Astrophysical Observatory of the Smithsonian Institution and associated mainly with the name of Abbot, who in 1895 joined the observatory and was its Director from 1907 till 1944, have provided a major contribution in the study of the solar constant during the first half of the present century.

In the beginning of the fifties Nicolet and Johnson having accomplished a revaluation of the magnitude of the solar constant, obtained its new value equal to 1395 W/m^2, or 2.00 cal/cm^2 min. This value became generally accepted in recent years (the finally accepted standard is the solar constant evaluated by Nicolet, equal to 1382 ± 28 W/m^2 or 1.98 cal/cm^2 min). The introduction of the international pyrheliometric scale (IPS-1956) put an end to the existing discrepancy between the American and European pyrheliometric scales.

Starting from 1961 a group of the faculty of the A.A. Zhdanov Leningrad State University, undertook the complex work of balloon sounding of the atmosphere during daytime up to a height of nearly 30 km, which enabled them to obtain data on the vertical profiles of direct solar radiation flux. In the works of Kondrat'ev and Nikol'skii [26, 30, 31, 112, 115] the first experimental proof has been provided, that the value of the solar constant at 1395 W/m^2 is an overestimation. A review of these and other investigations led to the conclusion that the most reliable maximum value of the solar constant is 1395 W/m^2, or 1,943 cal/cm^2 min (for IPS-1956).

33   In the second half of the sixties, several groups of investigators in the US began to record total and spectral measurements of direct solar radiation with the help of various items of equipment installed on aircraft and balloons. First of all the faculty of the Eppley Laboratory (headed by Drummond) and the Jet Propulsion Laboratory of the California Institute of Technology entered into these investigations. In the investigations conducted by these laboratories [81], the specially designed 12-channel aircraft radiometer with light (wide-band glass as well as interference) filters, which made it possible to carry out measurements of the total flux of solar radiation and flux of radiations in individual regions of the spectrum, had been used.

Thekaekara [177–179], at the Goddard Space Flight Center of NASA, analyzed several series of aircraft measurements of direct solar radiation with the use of filter radiometers as well as several spectrometers (integrated equipment includes twelve instruments of various types). Similar aircraft measurements were also carried out by Arvesen from the Ames Research Center of NASA. Murcray and his collaborators at the University of Denver undertook balloon measurements of vertical profiles of the flux of direct solar radiation.

All the data from the above listed aircraft and balloon investigations confirmed the conclusion in regard to the overestimation of the so-far accepted value of the solar constant. Stuiver and Ellis, Labs and Neckel [119], Makarova and Kharitonov [39], all of whom had devoted themselves to the reconsideration of extraterrestrial spectral distribution of solar radiation (and consequently of the solar constant), on the basis of data mainly from ground spectral measurements have also arrived at the same conclusion.

In this way, the numerous works published during 1968–70 led to the conclusion that the solar constant is less than 1395 W/m^2, or 2 cal/cm^2 min. It is natural that there exist known divergences between the results of individual investigations, especially in the determination of extraterrestrial spectral distribution of solar radiation. That is why a careful comparison and analysis of data obtained by various authors is required so that the final result is highly reliable. Such a comparison has been undertaken in a number of new works during recent years.

Fröhlich [91] studied the results of measurement of the total and spectral solar constant, carried out during the last decade, with the help of equipment installed on high altitude aircraft, balloons and satellites. In 34 [91], a brief summary characterizing the contemporary states of the absolute pyrheliometric scale is given and a new scale for comparison of the solar constant (SCSC) has been proposed. This is based on an analysis of the results of international pyrheliometric comparison at Davos in 1970 and 1975. The original as well as reduced SCSC values of the solar constant obtained by various scientists are reproduced in Table I.3. The averaging of all the values on the SCSC gives $S_0 = 1375$ W/m^2. The averaging of original values gives $S_0 = 1360$ W/m^2 with a mean square deviation of 14 W/m^2. Hence, modification of results related to the use of the SCSC is necessary. Apparently, the absolute error in the average value S_0 according to the SCSC is 20 W/m^2. Taking into account the possible overestimation of all the aircraft data, Fröhlich [91] recommends the value of S_0 in the range 1368–1377 W/m^2 as the most probable.

35   On the basis of analysis of aircraft and balloon measurements of solar radiation, carried out by Thekaekara [177–179], NASA adopted a recommendation to use the value of the solar constant at 1353 ± 21 W/m^2 as the standard. The standard spectral distribution of energy in an extraterrestrial solar spectrum in the range of wavelengths 0.115–1000 μm, which has been obtained mainly from the measurement data, was obtained with the help of five different spectrometers installed in the *Convair-990* aircraft laboratory.

This distribution (see [179]) considerably differs from those suggested earlier. Thus, for example, the visible range of spectral data obtained by Johnson appears to be on the higher side by approximately 5%, and the data of Labs and Neckel leads to an underestimation by 10%. A more

Table I.3. Value of the solar constant S_0 according to various authors

Author	Base	Period of measurement	S_0 W/m²	
			SCSC	original
Kondrat'ev and Nikol'skii	Balloon	1962–1967	1376 ± 18	1356
Drummond et al.	Aircraft *B-57B*	1966–1967	1387 ± 17	1359
Duncan et al.	Aircraft *Convair-990*	1967	1377 ± 40	1349
Krueger	—do—	1967	1372 ± 24	1364
McNatt et al.	—do—	1967	1375 ± 30	1343
Drummond	Aircraft *X-15*	1967	1385 ± 14	1361
Drummond et al.	Aircraft *Convair-990*	1967–1968	1387 ± 19	1359
Murcray	Balloon	1967–1968	1373 ± 12	1339
Kendall	Aircraft *Convair-990*	1968	1373 ± 14	1370
Plamondon	*Mariner-6, -7*	1969	1362 ± 18	1352
Willson	Balloon	1969	1369 ± 11	1366
Hickey	*Nimbus-6*	1975	1387 ± 14	1390
			1369 ± 14[1]	—
Hickey, Willson and Kendall	Rocket *Aerobee*	1976	1367 ± 7[2]	—

[1] Correction with consideration of rocket data.
[2] Preliminary results.

significant divergence exists for the ultraviolet and infrared regions of a spectrum. Such considerable discrepancies may arise because of less accurate measurements but, on the other hand, the possibility of a variability of the solar constant cannot be ruled out. Moreover, the existence of such a variability is well established in the ultraviolet (wavelength less than 0.3 μm) and microwave (2–10 cm) regions of the spectrum.

The lack of sufficiently reliable data creates an immediate requirement of long-term measurements of the total and spectral solar constants from satellites. The radiometer for measurement of the earth's radiation budget (*RMERB*), installed in the *Nimbus-6* satellite, has 10-solar channels for measurement of total and spectral solar constants in the intervals 250–300, 280–350, 300–400, 350–450 and 400–500 nm, separated with the help of interference light filters. Designed at the Goddard Space Flight Center, the solar energy monitor (SEM) is a combination of a prismatic monochromator with quartz optics, and a pyrheliometer. This is the only equipment that provides adequate resolution in the 0.25–2.6 μm band of the solar spectrum (the spectral width of the slit varies from 0.0012 μm on the 0.25 μm wavelength and 0.01 μm on the 0.6 μm wavelength to 0.05 μm on the 2.6 μm wavelength).

It is under consideration to use an aircraft prototype of the SEM for regular measurements of solar radiation from the *U-2* aircraft (altitude about 20 km). For a study of solar constant variations, it is important to maintain the functioning of an equipment of the same type over 11–22 years. Therefore, the SEM has been designed with small overall dimensions and weight, and can be easily installed in various satellites. In the beginning, it is proposed to use an SEM on the *Landsat* satellite and 36 cosmic carrier for multipurpose use during periods of maximum solar activity.

Willson [191] described the active-band radiometer (ABR), developed at the Jet Propulsion Laboratory of the California Institute of Technology, intended for the measurement of absolute values of the total flux of direct solar radiation in the automatic regime. The active-band radiometer has a receiver of radiation of a conical shape with an absorption coefficient of 0.997 ± 0.001. The absolute error of measurement of radiation flux in the range 0–0.150 W/cm^2 is 0.104–0.316 mW/cm^2. The angle of the circular field of vision of the equipment is $5°$. The comparison carried out with the Angström pyrheliometer during 1968–1972 showed that the ABR scale is overrated by 2.2% as compared to the International Pyrheliometric Scale (IPS-1956), while the accuracy of the ABR scale is more than $\pm 0.3\%$.

With the help of the ABR two series of balloon measurements of the total flux of solar radiation at altitudes of 25 km (1968) and 36 km (1969) were carried out. The corrections for the effect of light filters, thickness of the atmospheric layer and the distance between the earth and the sun enabled determination of the value of the solar constant. In accordance with the more reliable data, obtained in 1969, the solar constant is equal to 1366 ± 7 W/m^2 which is in agreement with the balloon data of Murcray et al. (1377 W/m^2) obtained after corrections for the divergence in the ABR and IPS-1956 scales. The analysis of data on the extraterrestrial spectral distribution of solar radiation, from the viewpoint of its correspondence with obtained values of the solar constant, led to the conclusion that the distribution obtained by Labs and Neckel [119] should be considered as most reliable.

Neckel and Labs [138] initiated a detailed analysis of various measured and calculated data on the extraterrestrial energy distribution in the solar spectrum (intensity at the center of the disk, darkening toward the edges, and mean intensity of emission of the entire disk, discrete emission and the continuum) in order to compile tables of the most reliable values of the spectral solar constant. Only that data, having a sufficiently high internal consistency (discrepancy between observational data on various days and seasons does not exceed 1%), which agrees with the model calculations and provides an accurate correspondence with the results of measurement of the total solar constant, the most probable value of which is 1360 W/m² (1.95 cal/cm² min), has been accepted. Apparently, the error in the determination of the solar constant ranges 37 between 0.5–1%. In [138], data on spectral distribution of extraterrestrial solar radiation in the range of 0.2–2.0 μm is presented and the necessity of further modification of the obtained results has been pointed out.

On comparing extraterrestrial spectral distributions of solar radiation, obtained in various works, a most significant disparity has been noticed in the ultraviolet region of the spectrum, amounting to 20% or more. In view of this De Luisi [79] designed a double monochromator with a system of automatic focusing on the sun which was used over a period of 2 years for high-altitude measurements of spectral distribution of solar radiation in the interval of 298–400 nm with the aim of obtaining the extraterrestrial distribution by way of extrapolation and also the determination of the weakening of solar radiation due to dust and ozone. The instrument is provided with a 45° quartz prism and a plane diffraction grating of 65 × 76 mm (1200 lines/mm). The angle of vision is equal to 1.12×10^{-4} sr (4.5' more than the angular diameter of the sun). The instrument is so far calibrated only in relative units.

In [79] it is shown that the effect of the solar corona never exceeded 1%. All the results of measurement of extraterrestrial spectral distribution of solar radiation in the interval 298.18–399.96 nm at a spectral resolution of 0.2 nm, obtained by way of extrapolation have been normalized in relation to the flux of solar radiation in the interval 300–400 nm accepted as 10.2 mW/cm² corresponding to the results obtained earlier by Thekaekara.

The comparison of the extraterrestrial spectrum with the data of other ground and aircraft measurements carried out during recent years enabled us to obtain a fairly satisfactory agreement (discrepancy does not exceed 5%). Hence, on the basis of ground measurements the changes in the extraterrestrial spectral distribution of solar radiation in the ultraviolet range of the spectrum exceeding 5% can be established.

In view of the fact that results of fresh assessments of the solar constant carried out during the past years indicated significant variations [from 1458 ± 3.5 W/m² (Sitnick, 1967) to 1343 ± 1.9 W/m² (McNatt and Riley, 1968)]. Crommelynck [77] pointed out that establishing the correct value of the solar constant depends considerably on the adequacy of the pyrheliometric scale and reliability of the Stefan-Boltzmann constant. On the basis of detailed analysis of results of a comparison of pyrheliometers and consideration of the discrepancies in pyrheliometric scales found in [77], the value of the Stefan-Boltzmann constant, equal to $(5.6926 \pm 0.074) \times 10^{-8}$ W/m²K⁴), has been obtained. With consideration of this 8 value and all data on measurements of the solar constant, obtained after

1960, Crommelynck proposed the most probable value of the solar constant as 1366.7 ± 5.3 W/m².

Detailed analysis of results of measurement of the total spectral solar constant, continued for more than a year on the *Nimbus-6* meteorological satellite launched on June 12, 1975, has been taken up by Hickey et al. [86]. The measurements ware carried out with the help of a 22-channel RMERB, 10 channels of which were earmarked for determination of the solar constant and its possible variations. The solar constant sensors carried out one series of measurements in each revolution (at the intersection of the southern terminator) and started functioning on the 2nd of July. The entire RMERB complex was switched on at first for two days after every 2 days, but later it was continuously operated.

Since in the analysis of data from RMERB there appeared certain difficulties related with the calibration of the apparatus, on June 29, 1976 this apparatus was launched by a rocket with the purpose of checking the calibration. The launching showed that data from RMERB and the corresponding solar constant are overrated by 1.4%. This has been taken into account in the final processing of the obtained results.

Table I.4 contains the characteristics of solar channels of RMERB and obtained results. Channel one is a duplicate and is switched on only once a month for the control of drift in the sensitivity of channel 2. A change in sensitivity with time was observed for all channels, except for channels 1, 3 and 5, and apparently, is caused by the impact of strong ultraviolet radiation on the light filters (especially on fused silicon).

Table I.4. Solar channels of RMERB and results, obtained in July 1975

Number of channel	Wave band, μm	Type of light filter	Average solar constant, W/m²
1	0.18–3.8	Suprazil	1369
2	0.18–3.8	—do—	1369
3	0.20–50	None	1392
4	0.526–2.8	Colored glass	969.2
5	0.698–2.8	—do—	675.8
6	0.395–0.508	Interference	205.9
7	0.344–0.460	—do—	160.7
8	0.300–0.410	—do—	109.2
9	0.275–0.360	—do—	55.2
10	0.252–0.324	—do—	24.2

During the first six months, variations of the total solar constant according to the data of channel 3 did not exceed 0.25% for the minimum value of 1390 W/m² and maximum value of 1393.5 W/m² (with a correction for change in the distance between the earth and sun). The accuracy of measurement is about ± 0.025% of the solar constant. An analysis of measurement data showed that about half of the variability of the solar constant is due to the effect of errors in guiding the sun. The remaining variability of approximately 0.1% should be attributed to either real variations of the solar constant or unknown equipment effects.

The data for June 29, 1976, indicates a decrease in the solar constant to 1388–1389 W/m². The value of the solar constant on July 2, 1976, was lower by approximately 0.4% than that of July 2, 1975. According to the data of channel 1 the mean value is 1370.8 W/m² for an amplitude of variation equal to 0.37%. The consideration of solar radiation, not measured by the sensors of channel 1, gives the value of the solar constant as equal to 1386 W/m², which is approximately 0.3% lower than the value of the solar constant from the data of channel 3, but still significantly higher than the expected value of 1370 W/m².

An analysis of data on the total solar constant, measured with the help of an RMERB mounted on the *Nimbus-6* meteorological satellite, revealed that the measured values of the solar constant appear to be higher, by about 1.5%, than expected. This motivated Duncan et al. [154] to take up special calibration of the RMERB by launching it on a rocket with ground comparisons and calibrations in the open air and in a vacuum chamber prior to launching. The complex of rocket equipment for the measurement of the solar constant included a satellite pyrheliometer, band-type absolute radiometers *Packrad* (Kendall design) and two Willson radiometers (PR-IV).

As a result of ground tests it has been observed that the readings of the equipment of the rocket complex were in agreement within the limits of ± 0.5%. The launching of the equipment on an *Aerobee-170* rocket was accomplished from White Sands testing ground on June 29, 1976. The solar tracking system functioned successfully. After normal functioning in flight the equipment was landed successfully by parachute and retrieved. In Table I.5 the values of the total solar constant obtained with the help of various items of equipment and also those measured simultaneously with the help of the RMERB device installed on *Nimbus-6*, are presented.

Table I.5. Preliminary results of measurements of solar constant

Instrument	Solar constant, W/m²
PR-IV, channel A	1368
PR-IV, channel B	1368
Packrad	1364
Pyrheliometer	1369
Rocket RMERB, channel 3	1389
Satellite RMERB, channel 3	1389

It is seen that the value of the solar constant from the RMERB data ('integrated' channel 3) is 1.6% greater than the value measured on the absolute band radiometer. The unweighted mean value on data of four absolute radiometers is equal to 1367 W/m² and its error does not exceed ± 0.5%. In this case complete agreement between the readings of the rocket and satellite RMERB is observed. Hence, it may be calculated that the values of the solar constant measured from the *Nimbus-6* satellite should be reduced by 1.6%. The detected discrepancies also indicate that the results of ground comparisons and calibrations should not be carried over to the conditions of functioning of the equipment in space.

Hickey et al. [103] reviewed the measurement of the total and spectral solar constant carried out over two years (starting from July 2, 1975) with the help of an RMERB mounted on the *Nimbus-6* satellite. The preliminary data on the values of the solar constant for various 'solar channels' of the RMERB corresponding to the state in July 1975 are given below:

RMERB channel, μm	0.18–3.8	(0.20)–50	0.526–2.8
Solar constant, W/m²	1369	1392	969.2
RMERB channel, μm	0.698–2.8	0.395–0.508	0.344–0.460
Solar constant, W/m²	675.8	205.9	160.7
RMERB channel, μm	0.300–0.410	0.275–0.360	0.252–0.324
Solar constant, W/m²	109.2	55.2	24.3

An analysis of data of subsequent measurements led to the conclusion on the critical importance of the problem of absolute calibration. The consideration of data of special calibrating rocket measurements undertaken on June 29, 1976, led to a mean weighted value of the total solar constant equal to 1370 ± 1 W/m². Since the integrating channel of the satellite RMERB gives a value of the solar constant equal to 1389 W/m², it should be considered overrated by 1.6%. The solar constant for the interval 0.18–3.8 μm is overrated by 0.6%. Consideration of this correction leads to a modified value which is equal to 1361 W/m². The reasons for overrating of readings of RMERB are still unclear. It is possible that a significant factor of discrepancy is the difference between the spectral composition of extraterrestrial solar radiation and radiation of a laboratory simulation of the sun used for ground calibration.

Channel 1 (0.18–2.8 μm) of RMERB is duplicated. One of the identical channels was switched on only once in a month on three occasions which enabled a study of the decrease in sensitivity of the working channel amounting to 10.3% in 18 months. A drop in sensitivity was also observed for the remaining channels, except integrating channel 2, which did not have an optical system. The change in the sensitivity of channels of RMERB was taken into account while processing the data.

An analysis of readings of channel 1 during 1.5 years led to the conclusion that the variability of the total solar constant is less than $\pm 0.1\%$. A similar result was obtained from the old radiation channel for 1976, but in 1975 variations up to 0.25% were observed. Hence, it may be considered that on the basis of measurement data for 1.5 years the variability of the solar constant appears to be less than 0.2%. With the aim of solving the problem of RMERB calibration, a self-calibrating band sensor for channel 9 has been installed on the *Nimbus-7* satellite.

From the moment of launching of the *Nimbus-3* meteorological satellite in April 1969, continuous tracking of the changes in the intensity of ultraviolet solar radiation in the 110–300 nm band has begun. With the launching of the *Nimbus-4* satellite on April 8, 1970, a simultaneous measurement of ultraviolet radiation with the monitor of ultraviolet solar energy, and also measurement of the global distribution of ozone from the measurement data of backscattered ultraviolet radiation in the interval 255–340 nm (double monochromator) and for 380 nm (photometer) were accomplished [98–100]. For the same period data on the interplanetary magnetic field is also available.

Heath [98–100], while analyzing the variability of ultraviolet solar radiation with time, discovered 3 types of periodicities, namely an 11-year cycle, a 27-day period of rotation of the sun and short-term variabilities caused by solar flares. In [100] the periodicities of the first two types are discussed. A comparison of all available data for the period from 1964 to 1972 shows that the radiation of a wavelength shorter than 210 nm, which is responsible for the production of atomic oxygen, can vary by a factor of two during the period between the maximum and minimum of the 11-year cycle.

The study of the sun as an ultraviolet star enabled us to observe its highly significant variability. The maximum intensities of ultraviolet radiation are observed at the moments of passage of active regions 42 through the central meridian of the sun. In this there exist two major active regions, located approximately at a distance of 180° on solar longitudes, which determine the presence of disturbances in a period of 13–14 days. The typical amplitudes of variations of ultraviolet radiations on wavelengths of alpha, 175 and 290 nm lines of the Lyman series are 25, 5 and 1% respectively.

From the viewpoint of the theory of paleoclimate the question of the possibility of solar constant variations in the geologic past is of great interest. In this connection Ellsaesser [85] pointed out that one of the major consequences of the evolution theory of stars as applied to the sun is the conclusion that the luminosity of the sun has increased by 25–35% over the 4.7 billion years of its existence. Taking into account this conclusion Sagan and Mullen showed, that if the composition of the earth's atmosphere and the albedo of the earth are invariable, then the mean temperature of the earth's surface should have been less than 0 °C even 2.3 million years ago. Such a conclusion, however contradicts geologic data indicating the existence of liquid water 3.5 to 4.5 billion years ago. This enabled the aforementioned authors to make assumptions in regard to the more severe greenhouse effect of the earth's atmosphere in the geologic past which could preserve the temperature of the surface above zero with 70% of the present-day luminosity of the sun.

From the generally accepted theory of evolution of stars it follows that the luminosity of the sun varies only in periods of the order of a billion years. The stability of luminosity is determined by the stability of nuclear reactions responsible for the generation of energy radiated by the sun. The transformation of hydrogen into helium has provided a stable existence for the sun over a period of the order of ten billion years, from the moment of its formation. The aforementioned theory predicts the emission of neutrinos as a result of nuclear reaction in the core of the sun and the possibility of detection of solar neutrinos on the earth [181].

The measurements of such neutrinos taken up in past years gave negative results and led to the conclusion regarding low threshold values of the number of neutrinos placing in doubt the generally accepted theory of evolution of stars. On the other hand this indicates the possibility of variability in the luminosity of the sun over periods of time relatively shorter than billions of years. The reasons for variations in luminosity of the sun may be related to the instability of the process, mainly in its zones, namely, the corona (solar winds), chromosphere, photosphere, convective zone and the main core.

The variability of solar winds and the chromosphere is well known but is not related with anywhere near noticeable variations of energy. Insofar as the photosphere is concerned, the main reason for its variability is sunspots. The variations of energy caused by their variability are also small, although there may be a more significant indirect effect of sunspots on luminosity through their impact on the convective zone. If the corona, chromosphere and photosphere are characterized by a variety of short-term temporary variations, then for the convective zone there exist three characteristic scales for the period of variations: 1) an 11-year periodicity caused by global circulation in a convective zone, 2) a circulation periodicity of the order of 1–10 days, 3) a periodicity of the order of 20000 years, determined by thermal inertia (time for warming or cooling), of a convective zone in the case of a change of energy fluxes from inside.

The measurements of solar constant variations carried out till now led to contradictory conclusions on the variability in luminosity of the sun. Therefore, the estimates of probable variations of the solar constant, oriented on the data on paleoclimate, are of interest. The numerical modeling of climatic changes caused by the sun showed, that the mean global temperature of the earth's surface should change by 1–2 °C in the case of a change in the solar constant by 1%. Since for a time scale of the order of 10^6 years, a variability of temperature of not more than 5 K is characteristic, this means that a variation in the solar constant by about 5% is permissible. As has been mentioned earlier, according to the models of evolution of the sun, the luminosity of the sun should have increased approximately by 25–35% since its formation to the present time. In this case the earth should have been totally covered with ice in the past. However, the greenhouse effect of the atmosphere prevented the earth from freezing.

Ulrich [181] studied the model of the main core of the sun, from which the possibility of variations in luminosity of the order of $\pm 10\%$ over a period of the last billion years follows. From the standpoint of the aforementioned estimates this model should be considered unrealistic. However, small variations in luminosity may be considered probable, if the existence of a relationship between the varying magnetic field and convection in the convective zone of the sun is assumed.

The results of investigations of the solar constant and its variability discussed above urge us first of all to draw a conclusion about the necessity of further investigations. As far as the total solar constant is concerned, apparently the interval of values 1368–1377 W/m² with the average 1370 W/m² proposed by Fröhlich [91] should be considered as most probable. The data generalized by Smith and Gotlieb [165] (Table I.6) can serve as supporting information on the extraterrestrial spectral

44   Table I.6. Extra-atmospheric spectral distribution of solar radiation in the interval of 290–4000 nm

Long-wave interval, $(\lambda_n - \lambda_k)$ nm	Solar radiation, W/m²		$S_{0-\lambda}/S_0$ %
	$S_{0.1}$	$S_{0-\lambda}$	
1	2	3	4
290–300	5.24 (−2)	1.57 (+1)	01.156
300–310	5.18 (−2)	2.09 (+1)	01.538
310–320	6.35 (−2)	2.72 (+1)	02.005
320–330	7.81 (−2)	3.50 (+1)	02.580
340–350	8.94 (−2)	5.30 (+1)	03.902
350–360	9.49 (−2)	6.25 (+1)	04.601
360–370	10.51 (−2)	7.30 (+1)	05.375
370–380	10.40 (−2)	8.34 (+1)	06.141
380–390	9.45 (−2)	9.28 (+1)	06.836
390–400	11.34 (−2)	1.04 (+2)	07.672
400–410	16.31 (−2)	1.20 (+2)	08.873

(Contd.)

1	2	3	4
410–420	17.00 (−2)	1.37 (+2)	10.125
420–430	16.59 (−2)	1.54 (+2)	11.347
430–440	16.72 (−2)	1.71 (+2)	12.578
440–450	19.28 (−2)	1.90 (+2)	13.998
450–460	20.06 (−2)	2.10 (+2)	15.475
460–470	19.86 (−2)	2.30 (+2)	16.938
470–480	19.89 (−2)	2.50 (+2)	18.403
480–490	18.88 (−2)	2.69 (+2)	19.793
490–500	19.56 (−2)	2.88 (+2)	21.234
500–510	19.02 (−2)	3.07 (+2)	22.635
510–520	18.31 (−2)	3.26 (+2)	23.983
520–530	18.59 (−2)	3.44 (+2)	25.352
530–540	19.17 (−2)	3.63 (+2)	25.764
540–550	18.56 (−2)	3.82 (+2)	28.131
550–560	18.41 (−2)	4.00 (+2)	29.487
560–570	18.28 (−2)	4.19 (+2)	30.833
570–580	18.34 (−2)	4.37 (+2)	32.184
580–590	18.08 (−2)	3.55 (+2)	33.515
590–600	17.63 (−2)	4.73 (+2)	34.814
600–610	17.41 (−2)	4.90 (+2)	36.096
610–620	17.05 (−2)	5.07 (+2)	37.351
620–630	16.58 (−2)	5.24 (+2)	38.583
630–640	16.33 (−2)	5.40 (+2)	39.788
640–650	15.99 (−2)	5.56 (+2)	40.956
650–660	15.20 (−2)	5.71 (+2)	42.075
660–670	15.55 (−2)	5.87 (+2)	43.220
670–680	15.16 (−2)	6.02 (+2)	44.337
680–690	14.89 (−2)	6.17 (+2)	45.433
690–700	14.50 (−2)	6.31 (+2)	46.501
700–710	14.16 (−2)	6.46 (+2)	47.544
710–720	13.85 (−2)	6.59 (+2)	48.564
720–730	13.56 (−2)	6.73 (+2)	49.562
730–740	13.16 (−2)	6.86 (+2)	50.532
740–750	12.84 (−2)	6.99 (+2)	51.478
750–760	12.65 (−2)	7.12 (+2)	52.409
760–770	12.36 (−2)	7.24 (+2)	53.320
770–780	12.07 (−2)	7.36 (+2)	54.209
780–790	11.83 (−2)	7.48 (+2)	55.080
790–800	11.61 (−2)	7.59 (+2)	55.935
800–810	11.36 (−2)	7.71 (+2)	56.771
810–820	11.04 (−2)	7.82 (+2)	57.585
820–830	10.75 (−2)	7.93 (+2)	58.386
830–840	10.51 (−2)	8.03 (+2)	59.150
840–850	10.06 (−2)	8.13 (+2)	59.891
850–860	9.86 (−2)	8.23 (+2)	60.617
860–870	9.68 (−2)	8.33 (+2)	61.330
870–880	9.47 (−2)	8.42 (+2)	62.028
880–890	9.24 (−2)	8.51 (+2)	62.708
890–900	9.20 (−2)	8.61 (+2)	63.386

45

1	2	3	4
900–910	8.98 (−2)	8.70 (+2)	64.047
910–920	8.74 (−2)	8.78 (+2)	64.691
920–930	8.57 (−2)	8.87 (+2)	65.322
930–940	8.41 (−2)	8.95 (+2)	65.941
940–950	8.23 (−2)	9.04 (+2)	66.547
950–960	8.06 (−2)	9.12 (+2)	67.141
960–970	7.89 (−2)	9.20 (+2)	67.722
970–980	7.73 (−2)	9.27 (+2)	68.291
980–990	7.56 (−2)	9.35 (+2)	68.848
990–1000	7.39 (−2)	9.42 (+2)	69.392
1000–1100	6.82 (−2)	1.01 (+3)	74.417
1100–1200	5.58 (−2)	1.07 (+3)	78.530
1200–1300	4.64 (−2)	1.11 (+3)	81.943
1300–1400	3.85 (−2)	1.16 (+3)	84.777
1400–1500	3.23 (−2)	1.18 (+3)	87.154
1500–1600	2.67 (−2)	1.21 (+3)	89.118
1600–1700	2.14 (−2)	1.23 (+3)	90.697
1700–1800	1.75 (−2)	1.25 (+3)	91.983
1800–1900	1.44 (−2)	1.26 (+3)	93.042
1900–2000	1.20 (−2)	1.28 (+3)	93.223
2000–3000	5.53 (−3)	1.33 (+3)	97.998
3000–4000	1.53 (−3)	1.35 (+3)	99.125

Note: In the columns 2 and 3 in brackets the values of the exponent (power) of 10 are shown (for example −2 means 10^{-2}). In column 2, the values of the spectral radiation flux $S_{0.1}$ (for the 0.1 nm waveband), in column 3, the values of flux of total radiation for wavelengths 0 to λ, including the considered interval are shown. In column 4, the fraction of radiation flux in the interval 0–λ in relation to $S_0 = 1358$ W/m^2 is given.

46 distribution of solar radiation. Satellite measurements carried out by Heath [98–100] enable us to judge the reality of short-term changes in the solar constant in the ultraviolet region of the spectrum. According to measurement data of the total solar constant from the *Nimbus-6* satellite, changes exceeding 0.1% were not revealed, but these results cannot be considered as final in view of the difficulties of calibration.

3. THE RADIATION BUDGET OF THE EARTH

As already mentioned the new trend in the investigations of the radiation budget of the planet relates to increasing attention toward the problem of the solar constant and its possible variability (see Sect. II), which creates an increased interest in the problems of contemporary climatic changes. This determines the efforts toward increasing the accuracy of satellite measurements of the components of the radiation budget of the earth's surface-atmosphere system [102], and as well as discussion on the usefulness of measurements with the help of wide-angled sensors [173, 195]. The discussion on possibilities of assessment of the radiation budget of the planet on the basis of computations has invoked considerable interest [1, 88, 89, 120, 148, 150, 151, 157]. In a number of works the possibility of determination of the components of the radiation budget of the underlying surface on the basis of data from satellite measurements has been considered [24, 44, 80, 127–129, 155, 180]. Primary attention was paid, however, to a review of the measurements of components of the radiation budget of the planet from satellites. The comprehensive investigations were dedicated to the discussion on measurement results of outgoing radiation from satellites of the *Cosmos* and *Meteor* series [23–25, 27, 32, 35, 40, 42, 45, 113]. Numerous publications in the form of reviews and original works contain results of analyses of measurement data from American meteorological satellites [65, 70, 76, 83, 84, 95, 96, 103, 108, 111, 114, 130, 131, 136, 139, 147, 152, 153, 173–176, 183–187, 192–196]. The results of investigations carried out in the USSR are contained in the monographs by Malkevich [40] and Andrianov et al. [20].

In order to obtain reliable data on the components of the earth's radiation budget, a correct method of processing measurement results is necessary. In order to work out such a method Campbell and Von der Haar [74] carried out numerical modeling for the estimation of accuracy of the determination of mean monthly values of the radiation budget on the basis of results of measurements of the total flux of outgoing shortwave and long-wave radiations with the help of wide-angled (semispherical) sensors in the case of nonuniform space time distribution of data. The calculations have been carried out for two models of angular distribution of the intensity of shortwave radiations characterizing the conditions above land and sea. In the case of a long-wave radiation field the darkening effect toward the edges of the disk were considered. An ideal sensor, located at a fixed point and registering round-the-clock measurements has been taken as a standard for comparison. The readings of this sensor were compared with the data obtained with wide-angled sensors mounted on a system of three satellites.

The calculations under consideration showed that determination of mean monthly values of the fluxes of long-wave radiation does not pose any difficulty since they do not exhibit any significant daily trend. The processing of data of measurements of shortwave radiation is much more difficult, especially in the event of a lapse during the mid-day hours. One of the possibilities lies in the use of data of ground measurements on the daily trend of radiation, reflected by diffusion reflectors. Another approach consists in observing the optimal periodicity of satellite measurements and obtaining actual data for corresponding moments of time.

The results, obtained till now, of measurements of the earth's radiation budget do not appear to be adequate from the viewpoint of their time span, consideration of the daily course and the characteristics of spectral composition of outgoing radiation. The data on the radiation budget is essential first of all for investigations on climate which can be classified in the following 3 categories: 1) numerical modeling of climate (input data, validity, sensitivity estimates and parameterization); 2) comprehension of the fundamentals of physics and chemistry of the climate; 3) climatic diagnostics (statistical relationship between various parameters of climate, etc.).

For a solution of this problem, NASA and NOAA (USA) are developing an optimal satellite system for measurement of the earth's radiation budget (SMERB) [192] with the help of a system of satellites equipped with instruments for measurement of the components of the earth's radiation budget in time scales of the order of months and longer with a provision for various spatial resolutions: 250×250, 1000×1000 km (in the tropics), for the $10°$ latitudinal zone. The problem consists, particularly in obtaining data on the meridional equator-pole gradient of the radiation budget and the mean global values of the earth's radiation budget.

On the SMERB satellites, the wide-angled (visible disk of the earth), medium-angled ($10°$ about the direction to the earth's center) and narrow-angled ($3°$) scanning sensors would be installed (for processing the results of measurements with the help of all these sensors, the data on the angular distribution of outgoing radiation is of great importance).

48   The calculations showed [192], that the use of two (or even better, three) satellites appears to be optimal from the viewpoint of scanning of the earth's surface and the cost of SMERB. One of them should be placed in orbit with an angle of inclination of $50-56°$ and at an altitude of 600 km (data of this satellite will contain information about the daily variation of the lower and temperate latitudes), and others in orbit with a wide angle of inclination, which would especially enable the complete illumination of regions of higher latitudes.

The SMERB equipment has 8 channels and consists of two systems: 1) wide- and medium-angled sensors with channels of 0.2–5 and 0.2–50 μm and also a channel of the total solar constant, and 2) narrow-edged scanning sensors with channels of 0.2–5, 5–50 and 1.6 μm (scanning is accomplished in the plane perpendicular to the orbit with a spatial resolution of 44 km at the nadir).

The anticipated accuracy of measurements of shortwave (SW) and long-wave (LW) radiation involves (determination for one month):

Spatial resolution	Accuracy of measurement, W/m²	
	LW	SW
250 × 250 km	9.4	10.3
1000 × 1000 km	9.4	10.3
10° latitudinal zones	5.2	5.3
Equator-pole gradient	2 (ERB)*	
Global average	1.3 (ERB)*	

*Earth's radiation budget.

Functional meteorological satellites of the 3rd generation such as *Tiros-N* would serve in the capacity of polar orbiting satellites of the SMERB, and in an orbit with an angle of inclination of 56° the 'AEM' type satellite will function as a television monitor.

Harrison et al. [98] analyzed the completeness of the array of measurement data obtained with the help of SMERB, while they paid special attention to the study of the effect of variability of cloud cover on the mean monthly values of outgoing shortwave radiation (0.2–5.0 μm). The main reason, as indicated, for the use of medium-angled sensors in SMERB equipment is the correspondence between the spatial resolution of these transducers and the scale of the geographical coordinate network, used for the study of processes on a regional scale and also the conveni-
49 ence of transformation of such data in terms of fluxes of outgoing radiation at an altitude of 30 km.

Analysis of the completeness of the array of data led to the conclusion, that data from two solar-synchronized satellites provides the necessary space-time view only up to 30%, but addition of one satellite in the orbit with an angle of inclination of 80° increases this view up to 60–68%. The location of an additional satellite in the orbit with an angle of inclination of 56° proved to be more successful as the view increased to 84–88%. Numerical modeling of the array of data obtained from SMERB consisting of 3 satellites (angles of inclination in the orbit being, respectively, 98, 98 and 56°) showed, that in this case the determination of the mean monthly zonal outgoing shortwave radiation is achieved with an error of less than 3 W/m². For outgoing long-wave radiation this error is less than 2 W/m².

The representation of results of measurements for individual regions led to the conclusion that in the given case (independent of latitude, time of the year and geographical location) the errors in determination of the mean monthly reflected radiation do not exceed 5 W/m², which corresponds to less than 1%, in the calculation of albedo. The accuracy of determination of the outgoing shortwave radiation happens to be almost the same in the case of spatial averaging over 500×500 km or 250×250 km. This demonstrates a sufficiently reliable consideration of the effect of changes in cloud cover for the use of the considered SMERB.

Keeping in view the problems of radiation climatology, Campbell and Von der Haar [75] carried out numerical experiments with the aim of revealing the optimal set of orbits for satellite measurements of the earth's radiation budget. The study of systems with two or three satellites in circular orbits at an altitude of 7178 km led to the conclusion that in the case of two satellites a combination of orbits with angles of inclination of 80° and 50° appears to be optimal. The optimum selection of angles of inclination in the three orbits gives values equal to 80, 60 and 50°. These results have been obtained under the assumption that use is made of flat or spherical sensors, functioning during daylight hours and providing a spatial resolution equivalent to what has been achieved earlier with the help of scanning radiometers of the *Nimbus-3* satellite. For a more realistic representation of the fields of outgoing shortwave and long-wave radiations their daily course has been considered.

With the aim of solving the problem of optimal selection of orbits calculations were carried out for 23 satellites with various angles of inclination of orbits and times of intersection at the equator. The most complicated part of the problem is to provide an account of the daily course with a minimum array of data. Since the flux of outgoing radiation at the level of the satellite is the modulated value of the flux at the level of the upper boundary of the atmosphere, a scheme of transition to the level of the upper boundary of the atmosphere with a partial elimination of the effect of modulation is discussed in [75].

The assessment of errors in determining the components of the earth's radiation budget showed, that even in the case of a two-satellite system the accuracy is fully acceptable. For the set of measurement data of long-wave radiation for one month, there appears a systematic error of 1.5 W/m^2, whereas for shortwave radiation this error is equal to 4.0 W/m^2. These errors increase in the determination of zonal averages with a resolution of 7.5° in latitude and in the worst cases are 30 and 10 W/m^2 comprising on the average 6 and 3 W/m^2, respectively, for reflected and outgoing radiation. In the case of a three-satellite system, the average errors are reduced to 5.3 and 1.3 W/m^2.

The combined use of data of solar-synchronized and other types of satellites also enables us to obtain good results, but in these cases the maximum errors increase. In the case of only one satellite the selection of a solar-synchronized orbit far from the terminator is preferable, rather than the precessional orbits. The most significant source of error appears to be inadequacy in consideration of the daily course of the radiation budget. Therefore, for analysis of the reliability of the array of data for solar synchronized and precessional orbits, more comprehensive information about the patterns of their daily course are of great importance. The application of this scheme of transition permitted us to increase the spatial resolution of data on long-wave radiation, but failed in the case of shortwave radiation (due to the effect of equipment noise).

In spite of serious efforts to develop equipment for measurement of components of the earth's radiation budget and to modify the method of processing, the data of similar measurements has still not attained such a level of accuracy and reliability, as is necessary for the requirements of its application in meteorology. According to Yates [199] the only practicable prospect for a radical increase in accuracy of satellite actinometric measurements is a transition to the measurement of relative parameters such as the earth's albedo. In this connection, the various systems of satellite actinometric equipment proposed in the US are described in [199].

Raschke [152] discussed the preliminary developments (phase A) of a research program, carried out with the support of the European Space Agency (ESA), and the principal features of equipment planned for the ESA climatological satellite, a sun-earth observatory. The basic function of such a climatological satellite is to observe the components of the radiation budget of the earth's surface-atmosphere system (including the solar constant), as well as some most significant parameters of the atmosphere and underlying surface determining the formation of the radiation budget. The satellite is proposed to be placed in orbit with a 57.3° angle of inclination and a 1150 km altitude, which would provide a daily precession of the plane of orbit of about 4° and enable us to obtain data, characterizing the daily course of the radiation budget (equatorial quasi-period of radiation covers 45 days).

For the measurement of the solar constant a pyrheliometer and solar spectrometer in the 0.16–3.5μm waveband (resolving power higher than 0.01μm) would be used. A wide-angled pyranometer and spectropyranometer are proposed to be used for the measurement of total and spectral fluxes of outgoing radiation. The system of these items of equipment supplements the multichannel scanning (narrow angled) radiometer meant for measurement of angular distribution of the shortwave and long-wave radiations.

The complex of actinometric instrumentation is equipped with a system of on-board calibration on the sun, space and black body. Since for obtaining the necessary quantum of data on the radiation budget prolonged (10–20 years) simultaneous observations with the help of 2–3 satellites are required, the climatological satellite is considered as only experimental. In this the measurement of the earth's radiation budget in future on a broad program, based on international cooperation, is kept in view.

In the framework of the research program of the D-58 French satellite (*Castor*), project BIRAMIS is being worked out with the aim of developing the equipment for measurement of the components of the earth's radiation budget by means of extrasensitive accelerometers capable of recording the acceleration of an artificial earth satellite with an accuracy of 5×10^{-10} m/s^2 [82]. The measurements are based on the principle of recording the acceleration of a satellite caused by optical pressure of direct solar, outgoing shortwave (reflected) and long-wave (thermal) radiations. It is proposed to use satellites placed in a circular orbit at an altitude of 1000 km and an angle of inclination of 70°. Estimates showed, that for a surface-to-mass ratio of the satellite equal to 0.02 m^2/kg, the acceleration caused by direct solar radiation could be 91.3×10^{-9} m/s^2.

52   Duhamel and Marchal [82] developed the model of the field of outgoing shortwave radiation with the consideration of diffusion and reflection components the use of which led to the conclusion, that scattering of outgoing radiation generates acceleration varying in the limits of $(0-20) \times 10^{-9}$ m/s^2 (depending upon the conditions of cloud cover and time of day). The mirror component contributes up to 5% of the direct solar radiation. The acceleration on account of optical pressure, caused by outgoing long-wave radiation, varies within the limits of $(9-13) \times 10^{-9}$ m/s^2.

Estimates of parasitic disturbances induced by external and internal factors were also obtained. Among the external factors are the deceleration by the atmosphere $(2.7 \times 10^{-9}$ m/s$^2)$ and photoemission from the satellite surface $(0.327 \times 10^{-9}$ m/s$^2)$. The contribution of internal factors does not exceed $(1-3) \times 10^{-9}$ m/s^2. The results, thus obtained, confirm the possibility of revaluation of the outgoing radiation for the night with an accuracy of about 1% and measurements on the other side of the earth with an accuracy of 0.5%. The time constant of the equipment may vary within the limits of 20–40 seconds.

The necessity of prolonged tracking of the parameters of the environment on the basis of observational data from satellite requires careful control of the sensitivity of the equipment used for this purpose, since experience has shown that the characteristics of various components of optical equipment in the case of their prolonged functioning in space are subject to significant variability. In view of this, Guenther [95] considered the possibility of controlling the sensitivity of optical equipment in the example of a 7.5 year functioning of a spectrometer installed on the *Nimbus-4* satellite for the measurement of backscattered (outgoing) ultraviolet radiation (satellite was launched in April 1970).

The results of measurements by means of this 12-channel spectrometer (waveband 0.25–0.40 μm) were used for establishing the total ozone content and vertical profile of the ratio of the mixture above the level of its maximum (30–60 km layer). The comparison of revaluations of total ozone content and vertical profiles of ozone concentration with synchronized data of ground observations obtained with the help of Dobson spectrophotometers and also with data of rocket and balloon soundings, serves as a basis in controlling the sensitivity of the equipment. With this aim, the comparison of measured and calculated values of backscattered ultraviolet radiation was also conducted (on the basis of data of rocket measurements of the vertical profile of ozone).

53   The new spectrometer launched aboard the *'Atmospheric Explorer-5'* satellite in November 1975 performed the function of calibration. The comparison of data of two satellites for September 1976 revealed variations in the ratio of measured values of outgoing ultraviolet radiation in the limit of 0.835–1.082 (depending on the wavelength). The results of this comparison, however, should not be considered convincing (from the viewpoint of the characteristic of changes in sensitivity of the spectrometer located on *Nimbus-4*), since the comparison was carried out only one year after the launching of the second satellite.

3.1. Radiation budget. Von der Haar and Raschke [185] reviewed the investigations of the global radiation budget on the basis of measurement data of outgoing radiation from the American meteorological satellites during the sixties. The measurements were carried out with the help of instruments of two types: scanning radiometers with an angle of vision of $5°$ and sensors for measurement of hemispherical fluxes of outgoing radiation (in both cases semiconductor bolometers served as receivers of radiation). The relative accuracy of measurement was about 3%.

A comparison of averages over the year, of global values of the radiation budget and its components on the data for 1962–1966 and 1969–1970 (*Nimbus-3*) reveals an appreciable agreement in the results: the earth's albedo is 29–30%, and the outgoing radiation 238 W/m^2. The radiation budget (within the limits of accuracy of measurements) comes to 0, if the solar constant is assumed as 1365 W/m^2. The annual course of the global radiation budget is small: only a very weak tendency toward an increase in albedo and decrease in the outgoing radiation from December to May is observed. The comparison of mean annual measured, and calculated, meridional profiles of the components of the radiation budget revealed a significant overestimation of the calculated values of albedo in the tropics. The data on the difference of radiation budgets of the equator and poles indicates that in both hemispheres, the meridional gradient of the budget is minimum during the summer and maximum in autumn, while there is considerable year-to-year variability of the gradient, specially for the summer of the northern hemisphere during 1962–1966. The latter demonstrates the importance of prolonged measurements of the global radiation budget with the aim of tracking its variability.

The processing of measurement data carried out earlier, on the fluxes of outgoing radiation obtained with the help of wide-angled sensors, made it possible to draw the first global maps of distribution of the radiation budget of the earth-atmosphere system and its components. These maps, however, characterized the highly modulated spatial fields of the radiation budget. The functioning of the five-channel scanning radiometer (FSR) installed on the *Nimbus-3* meterological satellite over a period of approximately 5 months (from April 16 to August 15, October 3–17, 1969 and from January 21 to February 3, 1970) enabled us to derive more detailed planetary distributions of the radiation budget of the earth-atmosphere system and its components (spatial resolution of 500×500 km^2 in the region $20°$S–$20°$N and 250×250 km^2 in the latitudes above $40°$). The parameters of satellite orbit (perigee 1097 km, apogee 1143 km, angle of inclination $99.9°$) permitted the measurement of outgoing radiation fluxes at tropical and subtropical latitudes twice in a day (at the moments of local midday and midnight).

Raschke et al. [174] discussed in detail the specific features of the processing of data, related with the necessity of consideration of the selectivity of sensors, relationships of angular distribution of radiation and daily course of outgoing radiation. The albedo of the planet was determined from the readings of channel 5 of FSR, which is sensitive in the waveband of 0.2–4.8 μm, while the outgoing long-wave radiation was determined from the set of readings of infrared channels 1 to 4 (the data of channel 2 was taken into account with maximum weightage of transparency windows of the atmosphere). Channels 1–4 possess an on-board calibration system (control of sensitivity), whereas the sensitivity of channel 5 was checked from the measurement data above flat reliefs of the deserts of North Africa and Saudi Arabia. The random equipment errors of all the channels do not exceed 1–2%. For the measurements, data for a nadir angle of more than $45°$ and a zenith distance of the sun of not less than $80°$, has been selected for processing.

While realizing the scheme of transformation from measured values of intensity of fluxes of shortwave radiation, the typical angular distributions of intensity for ocean, land surface and snow cover were used. This scheme of transformation for long-wave radiation is based on the typical angular distribution of outgoing radiation on the basis of the data from *Nimbus-2* for five geographical regions.

The authors [174] obtained average values of the radiation budget and its components for different 2-week intervals, plotted global maps for individual intervals and time-latitudinal sections for the whole period of observations. The mean global albedo is 28.4% and outgoing radiation 241.5 W/m^2, which corresponds to an effective temperature of 255 K. The radiation budget of the planet, if determined with an accuracy of more

than 1% (± 2.8 W/m^2) and the solar constant assumed as 1365 W/m^2 min
is equal to zero. This data, in agreement with the earlier obtained results
55 of measurements by the wide-angled sensors, indicates that the earth is
warmer and darker than what was inferred from climatological calculations.

In view of the peculiarity of the earth's orbit the annual mean insolation of the northern hemisphere is more than that of the southern. The
solar radiation absorbed in the southern hemisphere is higher (247.8) than
in the northern hemisphere (240.8 W/m^2), as the albedo of the northern
hemisphere is higher (28.7%) than that of the southern (28.0%). The
occurrence of a slightly higher outgoing radiation also determines the
negative sign of the radiation budget of the northern hemisphere (1.4
W/m^2), differing from the southern hemisphere (7 W/m^2). If these conclusions are correct, then there should be a transfer of energy from the
southern hemisphere to the northern.

Only a weak (2–4%) annual course of mean global values of albedo
and outgoing radiation is indicated, but a significant variability of the
mean zonal component of the radiation budget in a year is noted at almost all latitudes. The main peculiarity of planetary fields of the radiation
budget and its components is the presence of a steep longitudinal gradient
(along with a predominant meridional gradient), which primarily reflects
the peculiarity of distribution over land and sea. The comparatively lower
albedo in the Antarctic in January (not more than 65%), than according
to data of the *Nimbus-2* satellite (70–80%) is noteworthy. The errors in
determination of albedo and outgoing long-wave radiation may reach up
to 5%. In order to further increase the reliability of results, simultaneous
data of measurement of angular distribution of outgoing and incoming
direct solar radiation is essential. For control of accuracy, the synchronized measurements of radiation fluxes from balloons and aircraft can
be of significant importance. Only accomplishment of these subsequent
stages for increasing the accuracy of results would permit us to use them
for investigations on changes in climate.

Ellis and Von der Haar [83] obtained on the basis of satellite measurements the components of the radiation budget of the earth's surface-atmosphere system for the period from 1964 to 1971 (most reliable data for
29 months of this period was selected) and the mean meridional profiles of
the radiation budget and its components, averaged over the latitudinal
belts for intervals of time equal to a month, one season, or a year. For a
global average of albedo with the consideration of data for higher latitudes, only those conditions were considered for which insolation exceeded 1 W/m^2. In such a case the albedo in the latitudinal zone above 65°
varies within the limits of 50–60%. Since the measurement data was
obtained with the use of narrow-angled scanning sensors as well as wide-

56 angled (hemispherical) systems, the spatial resolution varied in the limits of 50–110 km in the first case and 1280–2130 km in the second (depending on the altitude of the satellite).

The mean zonal distribution was obtained for the latitudinal zone of 85°S to 85°N with a step of 10° in latitude, but the resolving power of wide-angled sensors approaches 20° in latitude. The solar constant is assumed to be 1360 W/m². An analysis of possible errors of measurement led to the conclusion that errors in the determination of the albedo and outgoing long-wave radiation amount is ± 5%. The error in determination of the radiation budget varies in wide limits, not exceeding 10 W/m² for average conditions over a year at all latitudes. In [83] data on mean meridional profiles of the radiation budget R, outgoing long-wave radiation F, albedo A, absorbed (S_a) and reflected (S_r) radiations for all the months of the year, seasons and average conditions over a year is reproduced in the form of tables and graphs. The data related to average conditions over a year is presented in Table I.7.

Table I.7. **Mean annual meridional profiles of radiation budget of the earth's surface-atmosphere system and its components (W/m²)**

Latitude	R	F	A	S_a	S_r
85°N	−103.2	174.7	58.9	71.5	102.4
75	−93.6	178.2	54.4	84.6	100.9
65	−72.1	189.1	45.2	117.0	96.5
55	−46.7	201.2	40.7	154.5	106.0
45	−20.9	2 8.3 [sic]	35.7	197.4	109.6
35	0.7	239.6	30.9	240.3	107.4
25	18.2	258.5	27.2	276.7	103.4
15	45.5	257.1	24.8	302.6	99.8
5	58.9	250.0	25.4	308.9	105.2
5°S	56.1	258.2	24.1	314.3	99.8
15	40.7	266.7	23.6	307.4	95.0
25	22.0	262.7	25.1	284.7	95.4
35	0.4	244.4	29.6	244.8	102.9
45	−27.3	224.4	35.8	197.1	109.9
55	−57.4	206.9	42.6	149.5	111.0
65	−85.6	189.6	51.3	104.0	109.6
75	−89.5	163.3	60.2	73.8	111.7
85	−87.7	154.3	61.7	66.6	107.3

Ellis [84] carried out a detailed analysis of relationships of the radiation budget and its components with the conditions of cloudiness by calculating the differences for actual conditions of cloudiness and clear 57 sky using the same array of data as in [83] (in the calculations of absorbed solar radiation the value of the solar constant was taken as

52

1360 W/m²). Since the results of measurements with the help of wide-angled sensors (characterized by a horizontal averaging of about 2000 km) constitute the major portion of the array of data under consideration, the results of measurements with the scanning radiometer of the *Nimbus-3* satellite for four fortnightly periods were used for assessment of the components of the radiation budget for a clear sky: May 1–15, July 15–31, October 3–17, 1969, and January 21– February 3, 1970.

The data, characterizing the mean annual meridional profiles of observed albedo and albedo in the absence of cloudiness, determined from the minimum values of albedo of the system, is presented in Fig. I.5. As is seen, the albedo of the system in a cloud-free atmosphere in the northern hemisphere is higher than in the southern. The mean minimum values of albedo in the latitudinal zones of 0–65° of the northern and southern hemispheres are 17.2 and 13.7%, respectively, while the quasi-global averages (65°S–65°N) is equal to 15.5%. For the entire globe (90°S–90°N) the mean minimum albedo is 16.9%.

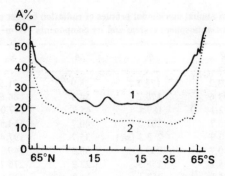

Fig. I.5. Mean annual meridional profiles of 1) observed albedo and 2) albedo in the absence of cloudiness.

The relationship between meridional profiles of outgoing long-wave radiation for actual conditions and for a cloud-free sky (maximum values) is characterized by the data of Fig. I.6. As in the preceding figure

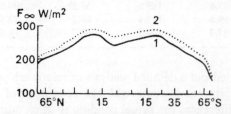

Fig. I.6. Mean annual meridional profiles of 1) observed and 2) maximum fluxes of outgoing long-wave radiation.

(maximum albedo is close to 7°N), here the effect of the ICZ (Intertropical Convergence Zone) (corresponding to the minimum of the outgoing radiation) is distinctly revealed. The mean global value of the maximum flux of outgoing radiation is equal to 261 W/m².

58   In order to analyze the relationship of the radiation budget of the system,

$$R = (1 - A) S_0 - F,$$

with the amount of cloudiness n, Ellis [84] calculated the difference

$$\Delta R = -(S_0 \Delta A + \Delta F),$$

where the difference between the albedo ΔA (S_0 being the solar constant) and outgoing radiation ΔF are determined by subtracting form the observed data corresponding values for a cloud-free atmosphere. The results of calculations of meridional profiles of mean annual values ΔR are presented in Fig. I.7. These results show, that except for the zones near the poles, it is always observed that $\Delta R \leqslant 0$. This indicates the predominance of a decrease in absorbed solar radiation, related to an increase in albedo due to increase in amount of cloudiness, over the decrease of outgoing radiation. The absolute values of ΔR increase in each hemisphere with increasing latitude approximately up to 60°. The value of ΔR varies from a minimum (7 W/m²) at 7°N to a maximum (49 W/m²) at 55°S. The asymmetry of distribution of ΔR over the hemispheres is mainly due to a nonuniformity of distribution of continental land.

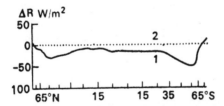

Fig. I.7. 1) Mean annual meridional profile of difference ΔR of observed radiation budget, and 2) radiation budget in the absence of cloudiness.

The meridional profiles ΔR above the ocean in both the hemispheres are practically symmetrical. In this case the effect of cloudiness on ΔR at higher altitudes of the sun, when the albedo of the underlying surface is minimum, becomes significant. The data characterizing the global and hemispherical effects of cloudiness is presented in Table I.8. In agreement with the results considered above, this table shows that, on the average over a year for the earth as a whole (more accurately for the latitudinal band 65°S–65°N), the effect of cloudiness on the absorbed radiation is stronger than on outgoing radiation, which determines

$\Delta R < 0$. The value of ΔR in the southern hemisphere is greater than that in the northern by 9.5 W/m², due to the difference in albedo of the underlying surface. The calculation of a global mean over the annual value for
59 the ocean gives $\Delta R = -30.6$ W/m². Hence, it is clear that the effects of cloudiness over land and ocean are significantly different.

Table I.8. Values of ΔR (W/m²) for both hemispheres and the earth as a whole, depending upon cloudiness

Latitude	Season				Year
	Mar-May	Jun-Aug	Sep-Nov	Dec-Feb	
65°S–65°N	−14.0	−6.0	−23.5	−26.7	−20.0
0–65°N	−14.2	−23.5	−2.0	−11.3	−15.3
0–65°S	−13.8	−8.5	−34.9	−42.0	−24.8

The processing of a complete array of data for 29 months resulted in the characteristics of the annual course of meridional profiles of albedo, outgoing radiation and radiation budget illustrated in Fig. I.8. As can be seen from Fig. I.8, the highest annual course is observed in the latitudinal zone of 30°S–30°N. The considerable increase of albedo in the 0–20°N zone, from 24% in May to 28% in August and the subsequent drop to 25% in October (Fig. I.8,a) merits special attention. This variability is the result of the effect of the southwest monsoon in Asia. A similar variability of the opposite sign, exists to the south of the equator. In the extratropical latitudes of both hemispheres, a considerable annual varia-
60 tion of the radiation budget and its components is observed. In extratropical latitudes of the northern hemisphere albedo attains a minimum value toward August-September, which is due to the minimum extent of snow and ice cover at this time of the year. At 45°N amplitude of the annual course of albedo attains 17%, whereas at 45°S it is only 7%. The outgoing radiation (Fig. I.8, b) exhibits an annual course which is a mirror image in relation to the variations in albedo. The data on the radiation budget (Fig. I.8, c) characterizes the degree of mutual compensation of the effects of variability of albedo and outgoing radiation. Such compensation precisely ensures the absence of an anomaly in the field of the radiation budget in the tropical zone. The distribution of the radiation budget, to a considerable extent is similar to the distribution of extraterrestrial insolation.

The estimation of the annual fluctuation of the global radiation budget also revealed a significant variability in this case. The data under consideration indicates the existence of a maximum positive earth radiation budget in March (16 W/m²) and a minimum in June (−14 W/m²). The

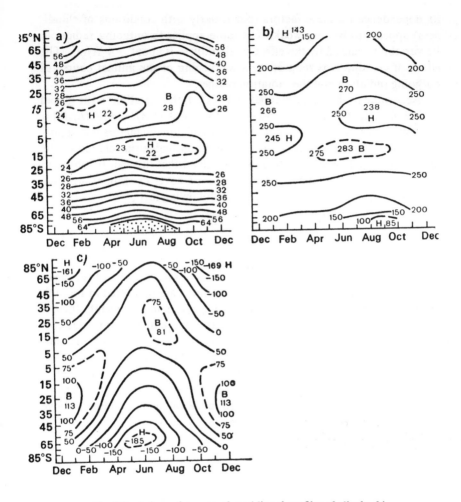

Fig. I.8. a) Annual course of meridional profiles of albedo, %,
b) outgoing radiation, W/m², and c) radiation budget, W/m²,
on the basis of data for 29 months (1964–1971).

amplitude of the annual course is reduced by a factor of two, if the effect
of eccentricity of the earth's orbit is not considerable, i.e. extraterrestrial
insolation is limited to the mean distance between the sun and the earth.

The earth's albedo, calculated by the application of reduced insolation, shows a double annual course with the amplitude of variation of
reflected radiation equal to 3 W/m², and minimums during the time of
equinoxes. The major factors determining the annual trend of albedo are
variations in the extent of snow and ice cover, cloudiness, vegetation
cover and the reflection coefficient of the underlying surface. However,

56

its dependence on these factors (particularly with conditions of cloudi-
61 ness) appears to be very weak. The same applies to outgoing radiation.
Its variability caused by the effect of cloudiness is about ± 4 W/m². The
effect of cloudiness is reduced to a weakening of the annual course of
outgoing radiation and the radiation budget.

According to the data obtained by Ellis [84] the variation in the global
radiation budget caused by observed cloudiness is $\Delta R = \Delta n \, (\Delta R/\Delta n) =
-33$ W/m². From the viewpoint of analysis of climatic factors it is impor-
tant to determine the variation in the radiation budget in relation to the
amount of cloudiness: $\delta = \partial R/\partial n$ (it is assumed that this can be obtained
only in respect of characteristics of the cloud cover existing at present).
The parameter δ can be presented in the following form

$$\delta = \frac{\partial S_a}{\partial n} = \frac{\partial F}{\partial n},$$

where $S_a = (1-A) \, S_0$ is the absorbed solar radiation. The results of computation for $\partial S_a/\partial n$ carried out on the basis of satellite data of the mean
meridional profile of the amount of cloudiness are presented in Fig. I.9.
If data relating to various seasons is also taken into account, then it
seems that the derivative of S_a varies within the limits from 0 (this corresponds to conditions of polar nights) to -128 W/m² at 58°S in Janu-
ary-February, and -110 W/m² near 58°N in July. On the average over
a year $\partial S_a/\partial n$ varies from -65 W/m² at 50°N to -86 W/m² near 58°S
(data for regions near the poles is not sufficiently reliable, which demonstrates the higher sensitivity of absorbed radiation to the cloudiness above
the oceans in the southern hemisphere.

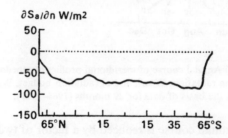

Fig. I.9. Mean annual meridional profile of sensitivity of absorbed solar radiation
to the amount of cloudiness.

The results of calculations of the sensitivity of outgoing radiation to
cloud cover $\partial F/\partial n$ are illustrated in Fig. I.10. The values of this deriva-
62 tive in the latitudinal band of 65°S–65°N are 1/3 to 2/3 of the derivative
$\partial S_a/\partial n$, which characterizes the measure of mutual compensation of
effects of shortwave and long-wave radiations. An increase in $\partial F/\partial n$ is

observed in the latitudinal band of 5–17°N with maximums near 14°N (data of Fig. I.9 does not reveal a similar anomaly in the case of absorbed radiation).

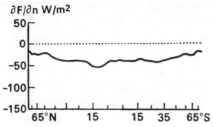

Fig. I.10. Mean annual meridional profile of sensitivity of outgoing radiation to cloud cover.

Over the entire 65°S–65°N belt $|\partial S_a/\partial n| > |\partial F/\partial n|$, which determines $\delta < 0$ (Fig. I.11), the mean annual values of δ vary from -42 W/m² at 55°N to -16 W/m² near 11°N, and then attain a value of -60 W/m² at 55°S. In Table I.9 estimates of mean global values of sensitivity of the radiation budget and its components to the amount of cloudiness according to data obtained by various authors are presented (for comparison the data corresponds to $n = 0.5$ and $S_0 = 1360$ W/m²).

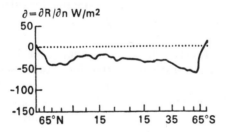

Fig. I.11. Mean annual meridional profile of sensitivity of radiation budget of cloudiness.

Table I.9. Mean global values of sensitivity of radiation budget and its components to the amount of cloudiness

Author	$\partial S_a/\partial n$	$\partial F/\partial n$	$\partial R/\partial n$
Ellis, array of data for 29 months (65°S–65°N)	-95.2	-53.1	-42.4
Ellis, data of *Nimbus-3*	-79.3	-41.6	-37.7
Adam (1967), 0–60°N	-178.1	-58.4	-119.7
Schneider (1972)	-129.2	-74.6	-54.6
Cess (1976)			
northern hemisphere	-88.4	-91	$+2.6$
southern hemisphere	-81.6	-81	-0.6

Since the calculations carried out by Cess were obtained with the use of modern satellite data, they should be highly comparable with the results obtained by Ellis. As can be seen from the tables, agreement between these results is limited, the closeness of the values of $\partial S_a/\partial n$. Much higher values of $\partial F/\partial n$ obtained by Cess, in practice, fully compensated the effects of shortwave and long-wave radiations, which is not in agreement with the data of Ellis and other authors. Such significant discre-
63 pancies show the necessity of further investigations on the sensitivity of the radiation budget and its components to the computations of cloudiness. No doubt, for average global conditions, such a situation is characteristic, when an increase (decrease) of the cloud cover with other factors being constant, causes an increase (or decrease) of the negative radiation budget of the system: a decrease in absorbed solar radiation with an increase in the cloud cover appearing to be greater than the decrease in outgoing radiation. From here it follows that the 'cloud cover-radiation budget' feedback mechanism is an important factor in the formation of climate.

Feddes and Liou [89] developed a method of calculation of total outgoing radiation on data for spectral channels of radiometers of very high resolution (RVHR) (located on the *Nimbus-6* meteorological satellite), bearing in mind the study of the relationship of outgoing radiation with conditions of cloudiness. The method is suitable for calculations of outgoing radiation in the case of single-layer clouds of middle or upper levels as well as multilevel cloudiness.

The 17-channel RVHR carries out scanning for 42 positions along every line of scanning (transverse to orbit) for spatial resolutions from 23 km at the nadir up to 31 km at the edge of the line. The channels of RHVR are the 4.3 μm and 15 μm bands of carbon dioxide, the 6.3 μm band of water vapor and the transparency windows (0.69; 3.68 and 11 μm).

In [89] only the data for those channels is considered for which the weightage function with the maximum below the hundred millibar level (data for the rest of the channels is practically independent of cloudiness) corresponds. The calculations of outgoing radiation for the model of a standard atmosphere, for summer at temperate latitudes have been carried out. It is assumed, that the lower boundary of clouds of the middle level and the upper boundary of clouds of the upper level are situated at the levels of 700 and 250 mb, and the thickness of the clouds varies in the ranges of 1.26–4.12 km and 0.36–2.68 km, respectively.

Calculations showed, that thick cirrus clouds considerably reduce outgoing radiation in the presence of both single-layer, as well as multi-layer cloudiness exerting a decisive effect on the formation of outgoing radiation. In the case of thick clouds at the middle level the outgoing radiation for all angles of sighting, exceeds the outgoing radiation in the case of thin clouds for channels in the carbon dioxide band. To these bands correspond the maximum of weightage functions, lying within the limits of thin clouds of the middle level or those above them.

Outgoing radiation is considerably more variable in the transparency window than in the carbon dioxide bands. It is also more sensitive to cloudiness in the shortwave channels of carbon dioxide bands than in the long-wave channels.

64   Feddes and Liou [89] analyzed, as an example for comparison of data obtained by RVHR with actual data, the possibility of revaluating the thickness of the cloud cover. There is a possibility of determining the thickness of cirrus clouds from measurement data of outgoing radiation for several infrared channels, with simultaneous use of images of cloud-cover in the visible and infrared regions of the spectrum, obtained from the *NOAA-4* satellite. The cloud thickness is estimated by comparing measured and calculated values of the ratio between outgoing radiation in the presence of clouds with that for a cloudless sky for various RVHR channels.

Starting from June 1974, scanning radiometers were regularly instal-led on *NOAA* meteorological satellites for measurement of the radiation budget of the earth and its components (outgoing long-wave radiation, albedo and absorbed solar radiation). On the basis of these measure-ments the mean monthly and seasonal maps of global distribution of the radiation budget are systematically drawn. An important feature of the data obtained is the fact that it pertains to 9 and 21 hours local time. Therefore, the earth's albedo is determined only for the morning hours, and the outgoing long-wave radiation is measured as the mean for data corresponding to two intervals.

Winston and Krueger [195] undertook an analysis of fields of the radiation budget and its components in connection with the studies of processes of formation and evolution of Asian summer monsoons on the basis of data for 1974–1976. The primary objective of the analysis was to reveal basic peculiarities of climatology of the radiation budget in the monsoonal belt and, particularly, to study the variability of the energy budget and monsoon from one year to another. Insofar as the Asian monsoon is a phenomenon on a semiglobal scale, its analysis is based on the data pertaining to the radiation budget over considerable sections of the eastern hemisphere and covering all the seasons of a year.

Results of observations made during the summer of 1975, immediate-ly reveal the effect of monsoonal circulation on the fields of cloudiness and outgoing long-wave radiation. The existence of thick clouds in the monsoonal zone is reflected in the radiation field as an extended region of outgoing radiation of less than 250 W/m^2. Vast deserts of North Africa

and Central Asia, where a clear sky predominates, correspond to the amounts of outgoing radiation exceeding 325 W/m^2. A relatively higher value is characteristic for the trade wind zone over the ocean. At all latitudes of the southern (during winter) hemisphere, considerable meridional gradients of the radiation budget are prevalent, whereas in the north-
65 ern (during summer) hemisphere the field of the radiation budget appears to be unclear. However, there exist significant zonal contrasts in the 20-40°N zone in the regions of the Sahara, Arabian Peninsula and the eastern regions of the Pacific Ocean with cooling in deserts and warming over oceans.

Apparently, a major portion of energy is conserved by oceans and is later released at another place and time. From the viewpoint of the formation of the Asian monsoon, the intense warming in continental Eurasia (north of 40°N) has greater significance than the surrounding oceans, this in turn determines the existence of a heat flow from the continent toward the ocean. The positive radiation budget in the region of Tibet goes up to 75-95 W/m^2. However, it still remains unclear up to what extent the warming of the continent due to a late spring and an early summer is responsible for the formation of the South-Asian monsoonal circulation.

Although, the differential warming of the land surface-ocean system undoubtedly, plays an important role in the development of the monsoon, the theory of the monsoon supported only on the consideration of the land-ocean system contrast appears to be inadequate. The consideration of warming due to condensation, in particular, and that of interaction of monsoonal and global circulation is essential. A study of variability over a year led to the conclusion that during a number of periods in the summer of 1975, the monsoon was more intense than during the summers of 1974 and 1976. This is confirmed by an increase in albedo and a decrease in outgoing radiation over extensive regions of India, the Indian Ocean, Southeast Asia and the Western section of the Pacific Ocean.

An important contribution in modulation of the radiation budget and atmospheric circulation comes from cloudiness, snow cover and other characteristics of the underlying surface. Particularly the increase in the extent of snow cover in Central Asia and the Himalayas resulting from winter and spring storms, served apparently, as a reason for the late start of the summer monsoon of 1976 (certain anomalies in far-off regions also contributed to this). In [195] the importance of a comprehensive analysis of the processes of monsoon formation taking into account not only satellite actinometric data, but also information on atmospheric dynamics and phase transformations of water (liberation of latent heat) in the atmosphere has been considered.

Winston [194] discussed in detail climatological patterns of variations in the earth's radiation budget on the basis of data obtained from a double-channel scanning radiometer used on *NOAA* satellites during June 1974 to March 1978. Since in the calculation of albedo the value of the solar constant was taken as 1353 W/m² and the temporal change in the sensitivity of radiometers was not taken into account, only the relative variability of components in the earth's radiation budget may be considered as sufficiently reliable. The absolute values could not be established.

66

A study of the data regarding the annual course of outgoing radiation and albedo for different years provides an explanation of the earlier established reflectivity of the annual course of these parameters: during summer, in the northern hemisphere, the minimum of albedo corresponds to the maximum of outgoing radiation, but during winter the reverse holds true. The hemispherical mean values, however, do not show such a relationship. In this case the annual course of outgoing radiation and the annual course of albedo are approximately in the same phase. The mean global annual course of outgoing radiation is determined mainly by a high annual variability in the northern hemisphere. An opposite situation is observed in the case of global albedo, in the annual course of which the high variability of albedo in the southern hemisphere and, particularly, a high albedo of the Antarctic plateau in summer provides a major contribution.

Year-to-year variations of global and hemispherical mean values are relatively small and are not always sufficiently reliable, but are more significant for individual regions. Earlier, for example, it was possible to establish the significant variability (on the basis of data for 1974–1977) associated with variations of cloudiness in the central and eastern parts of the tropical belt of the Pacific Ocean, cloudiness at lower levels in the eastern sectors of the Pacific and Atlantic oceans, and the snow and ice cover of Northern Canada at the beginning of summer.

A comparison of global maps of distribution of mean annual outgoing radiation for two years (June 1975–May 1976 and June 1976–May 1977) and an analysis of maps giving the difference between the outgoing radiation for the two years, revealed a similarity of fields of outgoing radiation for different years. However, there were certain significant discrepancies. A drop in the outgoing radiation (by more than 25 W/m²) in the central part of the equatorial belt of the Pacific Ocean, as well as a decrease of up to 20 W/m² in the eastern sector of the tropical Atlantic are clearly noticeable.

Apparently, the year-to-year variability under consideration exceeds the mean climatic variation, which was caused by stable extreme conditions of weather during 1976–1977 over a considerable part of the globe. The year-to-year variability of the earth's albedo and absorbed radiation are characterized by the presence of a microscalar structure, which is

67 probably related to the effect of variable cloudiness at the lower level. Such an effect is not noticeable in the case of outgoing radiation. In [194] the experience of drawing regional maps of the variability in absorbed radiation is also discussed.

The launching of the *Nimbus-6* meteorological satellite appeared to be an important stage in the investigation of the earth's radiation budget. The satellite was fitted with radiation budget equipment (RBE) which had ten channels for measurement of solar radiation in different spectral ranges as also over the integrated spectrum, four wide-angled (130°) sensors for measurement of long-wave, shortwave and integrated fluxes (0.2–50 μm) of outgoing radiation, and a scanning radiometer for the measurement of angular distribution of the intensity of outgoing radiation.

The RBE operated most satisfactorily. On-board calibration showed a high stability in the sensitivity of most RBE channels, except for a few solar channels. Serious problems, however arose in the processing of data of the wide-angled sensors, which were calibrated with an isotropic source (black body), whereas the earth is a source of nonisotropic radiation and does not completely fill the field of vision of the wide-angled sensors. For this reason discrepancies in the values of components of the radiation budget obtained from data of wide-angled and small-angled scanning sensors, emerged.

The equipment for measurement of the earth's radiation budget (ERB) aboard *Nimbus-6* and *Nimbus-7* satellites, has small-angled scanning detectors, which provide a spatial resolution of about 90×90 km at the nadir. The wavebands of 0.2–4 and 4–50 μm correspond to the channels under consideration of ERB equipment. This is aimed at measurement of ERB components in synoptic scales (approximately 500 km). Since for transition from measured intensities to hemispherical fluxes, information on angular distribution of the intensity of outgoing radiation is essential, the scheme of scanning considers the viewing of the same sections from different directions, which makes it possible to obtain empirical models of angular distribution for typical sections of the underlying surface (water, land, snow, clouds, mountains, etc.).

In order to build such models, the earth's surface is divided into 18630 sections, for each of which about 400 directions are fixed within the limits of the hemisphere [140]. Data of measurement, thus registered on magnetic tape contains information on the nature of the underlying surface of each section (radiation of dry land, water and snow or ice, topogra-
68 phy, etc.), cloud cover of lower, middle or upper levels and the solar altitude. Use of such an array of data enables us to construct an empirical model of angular distribution of the intensity of outgoing shortwave radiation for each section for given conditions of cloudiness and solar altitude.

The analysis of the results carried out by Jacobowitz et al. [140], reveals that the angular distribution of reflected radiation is most variable in the plane of scattering. For initial modeling of coefficients of reflection, the following typical underlying surfaces were selected: 1) ocean surface in a cloudiness atmosphere; 2) ocean surface in partial cloudiness; 3) land surface in the absence of clouds; 4) land surface in the presence of clouds; and 5) total cloud cover. After constructing the models of angular distribution of reflected radiation, the approximation of data obtained in the form of expansion in terms of empirical orthogonal functions is stipulated. In such a case the calculation of hemispherical fluxes can be carried out using the known coefficients of expansion.

The preliminary processing of ERB data, carried out by Smith et al. [139], gives a value of the total solar constant equal to 1392 W/m² which exceeds the anticipated value (1370 W/m²) by 1.6%. Both these values are related to the radiometric scale determined by the *Packrad* band radiometer, which is higher than IPS-1956 by 2%. The variability of the solar constant, with time, was less than 1%. On the basis of the data for July and August 1975, the mean global values of the earth's albedo, outgoing long-wave radiation and radiation budget are 30%, 240 and −4 W/m², respectively, which are in accordance with the data from the *Nimbus-3* meteorological satellite.

In [139] maps of global distribution of the radiation budget and its components according to the data of wide-angled sensors obtained during August 1975, are reproduced. The maps of the radiation budget show that cooling is taking place over almost the entire southern hemisphere (radiation budget is negative). The maximum positive values of the radiation budget are observed over the Caribbean Sea, in the western sector of the North Atlantic and eastern parts of the Pacific Ocean, whereas the minimum is observed only over North Africa.

A comparison with global maps, plotted on the data of small-angled scanning sensors, confirms an underestimation of the maximum albedo for Greenland and the Sahara from the data of wide-angled sensors. For this reason even the sign of the radiation budget of the Sahara according to data of wide-angled sensors appears to be erroneous (+20 W/m²), while scanning channels indicate a negative radiation budget from −20 to −40 W/m².

Smith et al. [176] summed up the measurements of components of the earth's radiation budget over one year commencing from July 1975 till July 1976, carried out with the help of wide-angled sensors of shortwave (wavelengths 0.2–3.8 μm) and integrated (wavelengths greater than 0.3 μm) outgoing radiation, mounted on the *Nimbus-6* meteorological satellite.

On the basis of data from indirect measurements, the values of albedo and outgoing long-wave radiation were calculated. The mean annual global values of these parameters are 31% and 236.3 W/m², respectively. The effective temperature of the earth is equal to 254 K. An analysis of the annual course of components of the earth's radiation budget on the whole led to the conclusion, that the annual course of absorbed solar radiation and the radiation budget are almost independent of the annual course of cloudiness (in all these cases the maximum is attained during the period December to February), and is mainly determined by the variability of extraterrestrial insolation, caused by the fluctuation in the distance between the earth and the sun.

The annual course of the albedo of the planet (determined by the variability of cloud cover and albedo of the underlying surface) causes only a small decrease in the amplitude of the annual variation of absorbed radiation. During a year, the total global outgoing radiation changes were very limited, which proves that complete mutual compensatory (for the entire earth) changes in its short and long-wave components under the influence of clouds take place. The regional values of the radiation budget of the earth surface-atmosphere system significantly varies in accordance with cloudiness and this, to a certain degree, influences the year-to-year variability of the global radiation budget. Such is the nature of difference between the earth's negative radiation budgets for July 1975 and July 1976. In the second case the absolute value of the budget is almost double. The year-to-year variability of the global radiation balance is determined mainly by differences in conditions in the subtropical belt of the northern hemisphere.

Analyses of meridional profiles of the radiation budget components for July during 1975 and 1976 reveal a considerable increase in long-wave outgoing radiation in the tropical and subtropical belts of the northern hemisphere in 1976, but a relatively lower variation in albedo. Apparently, cloudiness at the upper level was considerably lower in July 1976 (thick convective or cirrus clouds) than in July 1975. This variability is concentrated mainly above the continents and predominantly in the zones of Africa, the Middle East and Asia. Thus, regional variations in cloudiness can give rise to a significant year-to-year variability in the annual course of the global radiation budget.

The mean annual radiation budget of the earth stands at −2.6 W/m², if the solar constant is assumed to be 1391 W/m², but it becomes zero if the solar constant is reduced by 0.7% (to 1381 W/m²). Since the measurement accuracy of the ERB is 1%, it may be considered that the earth's mean annual radiation budget is zero within the limits of accuracy of measurements. An analysis of latitude-time sections of the radiation budget and its components show that during winter the albedo is maximum and outgoing long-wave radiation is minimum for each hemisphere (independent of latitude). This is due to an increase of cloudiness during winter, the extent of snow and ice cover and the drop in temperature of the underlying surface. In the tropics, the space-time variability of albedo and outgoing long-wave radiation is determined by the dynamics of the intertropical convergence zone (ICZ).

The observations reveal appreciable correspondence between the annual course of the radiation budget and variations in solar inclinations. The zone of the maximum (positive) radiation budget is located near 30°S in December, while most severe radiational cooling takes place in polar regions during winter and is sharply expressed above the Antarctic where the effective temperature drops to $-54°C$; (this is due to the high altitude of the Antarctic continent above sea level as compared to $-32°C$ in the Arctic.

Jacobowitz et al. [177] discussed the results of measurements of outgoing shortwave radiation (0.2–3.8 μm) and total radiation (wavelength greater than 0.3 μm) on the basis of data obtained over the first 25 months of the *Nimbus-6* satellite operation, with a comprehensive analysis of results obtained during the first 18 months of operation of the ERB apparatus. A comparison of the magnitude of fluxes of outgoing long-wave radiation determined from data obtained with small-angled long-wave and wide-angled integrated sensors (in this case the readings of shortwave sensors are subtracted) reveals that the latter appear to be systematically underrated. Since the small-angled sensors have a system of on-board calibration with respect to a black body, the measurements with wide-angled sensors were correlated with data from small-angled sensors for nighttime. In the daytime the data for shortwave wide-angled channels was corrected in such a way, that there was correspondence
71 between the values of outgoing long-wave radiation, determined as a difference of readings of integrated and shortwave channels and from the long-wave scanning data.

The comparison of data on the annual course of mean monthly values of albedo, outgoing radiation and radiation budget for two years (for the month of July in 1975 and 1976) reveals not only a relatively high repetition of the annual course, but also significant discrepancies. Considerable differences in albedo are observed, for instance, in October and November (in 1976, the albedo was higher than in 1975), whereas a divergence in outgoing radiation is noticed during February to July (in 1977, the outgoing radiation was more than in 1976).

As a rule, the annual course of albedo and outgoing radiation appear in opposite phases. In the second of the considered annual periods (1976), the radiation budget appears to be systematically lower in magnitude, especially over the period from February till June. The mean annual global values of the radiation budget for the first and second years are respectively -0.01 and -2.82 W/m².

The annual course of the radiation budget for the period from July 1975 to July 1976, is determined to a considerable degree, by the variability of extraterrestrial insolation. Thus, for example, in December 1975 there was a significant drop in the radiation budget caused by the increased total outgoing radiation. This is related with the rise in reflected radiation as a result of increase in albedo as well as insolation; moreover this last factor compensates a simultaneously occurring decrease in outgoing long-wave radiation.

An analysis of time-latitudinal variation of albedo, outgoing radiation and the radiation budget for 18 months reveals known maximums (minimums) of albedo (outgoing radiation) during winter months of the corresponding hemispheres at all latitudes. This is caused by the combined effect of an increase of clouds, extent of snow and ice cover and also a fall in the temperature of the underlying surface. The tropical belt remains an exception, where the variability of the ICZ exerts a decisive influence on the annual course of albedo and outgoing radiation.

The extraterrestrial insolation exerts a very great influence on the field of the radiation budget. For example, the occurrence of the maximum radiation budget of 120 W/m² in the southern hemisphere in December, where its maximum in the northern hemisphere in June-July amounts to only 80 W/m², confirms this statement. In both hemispheres a considerable increase in meridional flow of the radiation budget is observed during winter. The maximum positive budget occurs in December near 30°S, and maximum negative budget at higher latitudes (up to -180 W/m² in the northern hemisphere). The radiational temperature in the Antarctic drops to -54 °C and above the North Pole to -32 °C, during winter.

The application of measurement data on components of the earth's radiation budget for numerical modeling of climate and its measurement requires the representation of measured quantities as functions of those parameters which figure in the theory of climate. In view of this, Jenesius et al. [107] proposed a method for calculating outgoing long-wave radiation with the use of parameters which can be calculated beforehand with the help of numerical models used in weather forecasting. Such parameters have been selected on the basis that their relationship with the conditions of cloudiness are either known or assumed. This means, that the method under consideration implicitly takes the dependence of outgoing radiation on cloudiness into account and hence opens up the possibilities of obtaining an implicit model of cloud feedback mechanism.

The regression equation obtained from data for the winter 1974–1975, related to the territory of the USA, constitutes the basis for the method meant only for the continental winter condition. For derivation of a regression equation the measurement data on outgoing radiation in the waveband 10.5–12.6 μm from *NOAA-3* and *4* has been used as the source (it is known, that a correlation coefficient of outgoing radiation in this spectral interval with total outgoing radiation may take values of up to 99%).

The measurement data has been obtained for a spatial resolution of 10 km, but averaged corresponding to the pitch of the grid used in numerical forecasting for the USA (190.5 km at 60° latitude). The multiple linear regression was carried out with the help of the temperature of the boundary layer of the atmosphere (TBL) and two Planck functions being used for temperature prediction in the 500–1000 m layer. One of the Planck functions was determined by the TBL and the second by radiational temperature T_r which is a combination of TBL, air temperature and the dew point at the level of 500 mb. The predictions characterizing the distribution of humidity and wind have also been taken into account. The regression equation, thus obtained takes the form (W/m²):

$$F = 310.0 + 5569 \, T_r - 0.266 \, R_m - 9.30 \, J_{35} - 0.126 \, H_6 - 0.246 \, V_2,$$

where R_m is the maximum relative humidity of the air in one of the layers of a model of numerical forecast; J_{35} the wind velocity on the anticyclonic side of the jet flow (35 m/s or more); H_6 the variation in the altitude of the 100 mb surface over 6 hours; and V_2 the wind velocity at the level of 200 mb. The contribution of all these components of variability of outgoing radiation is approximately the same except for T_r, the weightage of which is approximately 5–7 times more than any of the remaining components. The consideration of T_r explains 60.8% of the daily variability in relation to the 3-month average values, whereas all remaining factors increase this value up to 67.3%.

The testing of an independent array of data for February 1976 demonstrated that the regression equation describes a 97.8% deviation of the mean values of outgoing radiation at grid points. The mean square deviation of calculated mean values from the observed is 6.6 W/m,² whereas the mean square deviation of observed mean values is 28.9 W/m². A very slight tendency toward overrating calculated values at higher latitudes and underestimating them at lower latitudes is observed. Computed daily outgoing radiation fields appear to be smoother than those observed, but they clearly reveal the relative minimums of outgoing radiation caused by clouds with a higher upper boundary. If a similar regression equation could be obtained for reflection shortwave radiation, this might permit computation of a radiation budget of the earth's surface-atmosphere system.

The use of measured components of the earth's radiation budget enables us to solve various problems of atmospheric energetics, which are of immense interest from the standpoint of the problems of global climate. In view of this, two directions of such investigations will be discussed:

1. study of atmospheric energetics and transfer of heat in the oceans; and

2. simulation of the components of the underlying surface radiation budget in accordance with the data of satellite measurments.

3.2. Atmospheric energetics and transfer of heat in the oceans. Investigations based on the use of results of satellite actinometric measurements showed that, on the average in the atmosphere and oceans, a meridional transfer of energy takes place over a year in comparable magnitudes. The ocean is primarily responsible for a transfer of energy in lower latitudes, and the atmosphere—in temperate and higher latitudes. Oort and Von der Haar [141] undertook to study this annual course of the components of meridional transfer in the northern hemisphere and consequently, the role of the oceans and the atmosphere as reservoirs of heat.

The results obtained in [141] are based only on satellite data and the usual meteorological and oceanographic observations, which include:

1. satellite data on reflected solar radiation and outgoing long-wave radiation during 1964–1971;

2. data of approximately 600 stations for aerological sounding in the
74 northern hemisphere during the period 1958–1963; and

3. all available results of oceanographic observations.

The thermal budgets between the atmosphere and land or the atmosphere and sea are examined separately. In the first case the following components are considered:

1. the shortwave budget at the level of the upper boundary of the atmosphere;

2. thermal reserves of the atmosphere;

3. influx of heat due to advection; and

4. heat flux through the underlying surface (the latter calculated as a residual term).

The 'earth component' of the thermal budget includes thermal reserves of the oceans, land surface and ice cover as well as the energy exchange on account of transfer in the oceans.

In order to determine the least studied component, i.e. heat flux in the oceans, the following indirect methods have been used:

1. estimation of localized convergence of heat by ocean currents, as a remainder term of the thermal budget equation of the atmosphere-ocean-earth system;

2. similar estimation by using the thermal budget equation of the ocean surface;

3. calculation of the steady part of the heat flux from the observational

data regarding mean temperature and computed geostrophic wind field (the latter is determined from the three-dimensional density distribution); and

4. use of data of numerical experiments with a very small pitch of the spatial grid to estimate the heat fluxes, as well as the relative role of the mean current and vortexes in the thermal budget of the ocean.

An analysis of obtained results (average monthly meridional profiles of thermal budget components) confirms the conclusion arrived at earlier regarding heat reserves in the ocean dominating the total heat reserves of the atmosphere-ocean-land surface-cryosphere system. Maximum variations in the heat reserves of the ocean are observed in the tropical belt of 10–20°N with a maximum during spring and a minimum during late summer. An important discovery is that, for a meridional transfer of heat in the ocean, a marked annual course especially in the tropics, is characteristic.

Calculations for the 10–20°N belt gave values of heat flux (through the latitudinal circle during spring and late autumn) equal to $(4-5) \times 10^{15}$ W. This is comparable with, or exceeds the corresponding heat flux at temperate latitudes. Close to the equator during a year, a change in the sign of heat transfer takes place which as a rule is directed toward the hemisphere undergoing winter for an absolute maximum of 8×10^{15} W in August.

A typical situation reveals, that at higher latitudes an equilibrium between the heat loss, determined by the negative radiation budget of the system, and a convergence of heat caused by atmospheric movements, is observed. In the tropics the positive radiation budget of the system is
75 counterbalanced by a divergence of heat under the impact of sea currents. A considerably more complicated relationship of the components of the thermal budget is noted at temperate latitudes where it is essential to take into account effects due to all the components. The significant conclusion of [141] is that oceans play an extremely important role in the determination of climate. Therefore, direct measurements of heat advection in the ocean by following the corresponding international program appears to be a very important task.

In view of the application of satellite data regarding the earth's radiation budget in the investigations of atmospheric energetics, it is important to discuss certain general aspects related to this problem. In order to establish the adequacy of the models of climate and general atmospheric circulation it is essential to know not only the observed statistics of meridional parameters, but also the degree of fulfillment of various physical laws of conservation (energy, momentum, etc.) in the model atmosphere. In diagnostic investigations of this type carried out so far, as a rule, only the axisymmetrical conditions were considered. However, the problem consists in clarification as to whether the laws of conservation hold good in different parts of the globe. In this connection, Holopainen [104] undertook investigation of the geographical distribution of certain components of the vortex and kinetic energy budgets, especially of the effect of the so-called large-scale turbulence in the time-averaged flow.

The data on statistics of wind in the free atmosphere of the northern hemisphere for the two independent five-year periods (1950–1954 and 1958–1962) serves as an empirical basis for study. All the values considered here are averaged over a year for the 100–1000 mb layer. Analysis of global maps of the relative vortexes for both the periods reveals a similar picture and indicates the predominance of an anticyclonic vorticity at lower latitudes and a cyclonic vorticity at temperate and higher latitudes. Large differences between the two sets of data (approximately by two times) are found in the northern region of Siberia and over the Pacific Ocean, which apparently is caused by the insufficient duration of the series and inadequacy in the methods of data processing. These discrepancies emphasize the large gaps in our understanding of the general circulation of the atmosphere.

A comparison of absolute vorticity in well illuminated regions of North America and Europe for these two five-year periods enabled us to observe only small differences. On the basis of the vortex equation in [104], the factor determining the variations in vorticity fields, viz. horizontal advection of mean vorticity by average wind, the effect of the Reynold stresses, mean divergence, frictional forces, etc., have been analyzed. First of all, the effect of a three-dimensional distribution of Reynold stresses $\overline{u'u'}$, $\overline{u'v'}$ and $\overline{v'v'}$ (u', v' are fluctuations of the horizontal components of velocity) on the averaged current has been analyzed.

Calculations demonstrated that it is important to take into account the effect of velocity fluctuations in the study of the time average of resultant vortex and, consequently, in the determination of the time averages of the vertical component of the velocity field. An investigation of the evolution of kinetic energy of large scale turbulence as a function of latitudes and longitudes led to the conclusion that the geographical distribution of kinetic energy in both the periods considered appeared to be similar: there are distinctly expressed maximums above oceans and minimums above continents. Analysis of the results of computations relating to convergence of the horizontal component of the kinetic energy flux (caused by moderate movement) and that of the vertical component of the energy flux shows, that as a rule the latter proves to be negative. Hence the phenomenon of negative viscosity.

In [105] Holopainen reviewed the contemporary status of the problem of the global energy budget, which is of prime importance for developing an adequate theory of climate. In this connection, first of all a few classical concepts, relating to the problem of the mean annual thermal budget of the earth, have been considered. Attention is also drawn to those components, the contemporary quantitative estimates of which differ considerably from those obtained a few decades earlier.

As already mentioned, analysis of data of setellite measurements showed that the earth on the whole and particularly the region of tropical latitudes, is darker (its albedo is lower) than was assumed earlier. The year-to-year variability of global albedo and cloudiness remains practically unstudied. The earth's outgoing thermal radiation appeared to be considerably higher than the value obtained earlier from the computational data.

In Fig. I.12 a schematic diagram of the mean annual thermal budget of the earth (in relative units) based on the data of Schneider and Dennett [161] is given. According to this, the global albedo is equal to 28%, while clouds bring in a maximum contribution to the albedo (19%). The contribution by atmospheric molecular and aerosol scattering under cloudless conditions stand at 6% and of the underlying surface at 3%. In the absorption of shortwave radiation (72 units) 25 units are accounted for as the contribution of the atmosphere and 47 as the share of the underlying surface. The atmosphere absorbs almost all thermal radiation of the underlying surface (109 and 114 units). In the presence of such a strong greenhouse effect, the outgoing long-wave radiation is determined by the radiation of the upper boundary of the cloud cover and upper layers of the atmosphere (CO_2 and H_2O). The radiation budget of the atmosphere is negative (-29 units) and that of the underlying surface is positive but has the same absolute value. The transfer of energy from the underlying surface to the atmosphere is chiefly determined by the latent heat of condensation (24 units) and only to a small degree by a turbulent heat exchange (5 units).

From the viewpoint of the physics of climate, the consideration of redistribution of energy in a horizontal direction is of great importance. In this connection, Holopainen [105] considered the equation of the thermal budget of the underlying surface, the atmosphere and underlying surface-atmosphere system for the vertical column of the atmosphere.

A schematic diagram of the local energy budget of the atmosphere-ocean-land surface-cryosphere system is presented in Fig. I.13. The equation of energy budget of the system takes the form:

$$S_A + S_O + S_L + S_I = F_{TA} - \text{div } T_A - \text{div } T_O,$$

where $S_i (i = A, O, L, I)$ is the variation of total energy of the corresponding mediums; F_{TA} the nonconvergent flux of energy at the level of the upper boundary of the atmosphere (radiation budget); F_{BA} the vertical

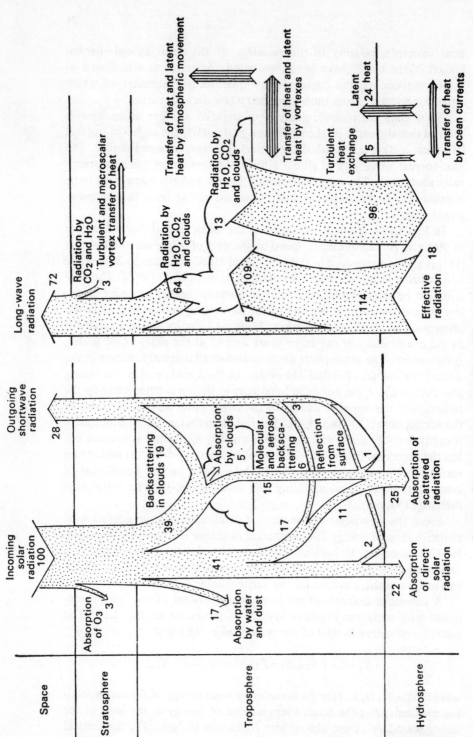

Fig. I.12. Scheme of mean annual thermal budget of the earth.

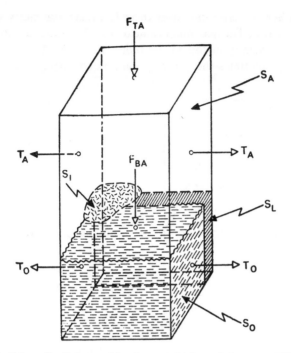

80 Fig. I.13. Schematic diagram of local energy budget of the ocean (O)–land surface
(L)–snow and ice cover (I)–atmosphere (A) system.

flux of energy at the level of the underlying surface, and T_i $(i = A, O)$ the
horizontal transfer of energy in the atmosphere and ocean.

Analysis of the global distribution of various components of the thermal budget demonstrates that they have not been fully understood. This
applies specially to horizontal transfer of heat in the oceans. If we consider that the distribution of the radiation budget of the system and total
energy influx to the underlying surface (determined by the radiation
budget, turbulent heat exchange and heat exchange due to phase transformation of water) are known and the averaging over latitudes and time
(for a year) is realized, it is possible to determine the required meridional
fluxes of energy in the atmosphere (T_A) and in the ocean (T_O). The values
found by direct calculations from data on three-dimensional fields of
corresponding parameters are the same as the observed values of such
parameters.

In [105] a method of computation of the required energy flux in the
ocean is described, the accuracy of which primarily depends on the reliability of determination of the total energy influx to the underlying surface.
The calculations carried out by Palmen and Newton (Fig. I.14) showed
that the total meridional flux $(T_A + T_O)$ of energy is directed toward the

poles in both hemispheres; moreover there is a quasi-symmetry with re-
78 spect to the equator. The maximum value of $T_A + T_O$ (about 5×10^{15} W) is
observed in the vicinity of the 35° latitude. Interestingly, these values
exceed by only 500 times the contemporary global energy consumption
(10^{13} W).

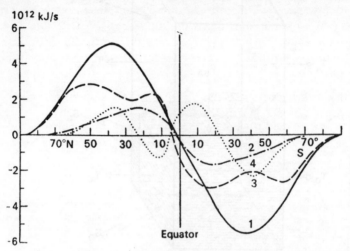

Fig. I.14. Mean annual values of 'required' meridional fluxes
of energy toward poles:
1—total energy flux $(T_O + T_A)$; 2—T_O; 3—energy flux due to latent heat (T'_A)
and 4—dynamic energy flux (T''_A), $T_A = T'_A + T''_A$.

There is significant meridional transfer of energy in the oceans, which
is almost half of the total energy flux at subtropical latitudes (10–20°).
It must be emphasized that the evaluation of the required T_O is only
approximate. As far as transfer of energy in the atmosphere is concerned,
a significant portion of it in both the hemisphere constitutes the latent
79 heat of condensation from the subtropical zones toward the poles, as well
as toward the equator. The remaining portion of T_A is the 'dynamic'
energy. For meridional profiles of the required T_A, the presence of two
maximums at lower and temperate latitudes is characteristic.

A comparison of values of T_A with 'observed' values computed from
the aerological data, indicates a satisfactory agreement. However, for
T_O such a comparison is still not possible. Analysis of the observed T_A
has established that the transfer of all forms of energy in the atmosphere
81 at lower latitudes is determined mainly by the effect of meridional circulation, whereas at temperate and higher latitudes it is determined by the
effect of large scale vortexes.

Von der Haar and Oort [184] used the data of satellite observations
of the radiation budget F_{TA} of the system for fresh calculations of $T_A + T_O$

and T_O in the northern hemisphere, which led to considerable higher values of T_O. This appears to be a direct consequence of overestimation of the earlier used albedo of the system (Fig. I.15). The satellite data for 1964–1971 enabled a study of some peculiarities of the annual trend of components of the global energy budget. Oort and Von der Haar [141] found, for example, that the entire northern hemisphere is warmed during summer ($F_{TA} > 0$) and cools down during winter ($F_{TA} < 0$). Figure I.16 indicates the data obtained in [141], characterizing the annual course of meridional profiles of the radiation budget of the system in the northern hemisphere.

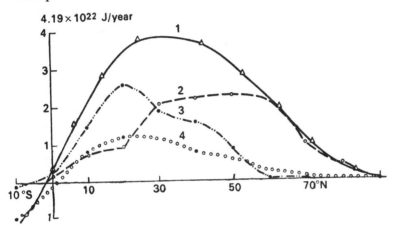

Fig. I.15. Meridional transfer of energy in northern hemisphere, determined by taking into account satellite data of measurements of radiation budget: 1—'required' transfer of total energy ($T_O + T_A$), calculated from the data of satellite measurements of radiation budget of the system (F_{TA}); 2—'observed' transfer of energy in the atmosphere (T_A); 3—'required' transfer of energy in oceans (T_O); 4—earlier estimates of 'required' transfer of energy.

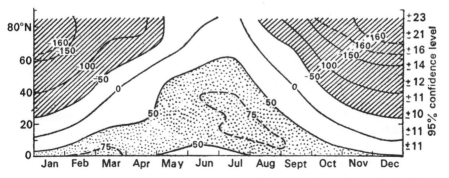

Fig. I.16. Annual course of meridional profile of earth's radiation budget in northern hemisphere according to data of satellite measurements during 1964–1971.

The total energy of the atmosphere (sum of internal, potential and kinetic energies) increases (variation of total energy $S_A > 0$) from the beginning of February till the end of July and decreases during the rest of the year, and the maximum amplitude of S_A occurs at higher latitudes. For the observed energy flux T_A in the atmosphere, a fairly regular annual
82 variation (with maximum energy flux toward the poles during cold periods of the year and minimum during warm periods of the year) is characteristic.

There is a small transfer of energy through the equator from the summer to the winter hemisphere. The required energy flux T_O into the ocean is directed, as a rule, toward the pole and it is comparable with the values of T_A. A strong annual course of T_O at lower and temperate latitudes is observed. Apparently, there is an intense transfer of energy in the oceans across the equator. A comparison of T_A and T_O leads to the conclusion that in meridional transfer of energy at lower and subtropical latitudes of the northern hemisphere, the influence of the ocean is dominant, whereas at temperate and higher latitudes the influence of the atmosphere dominates. These results establish the prime necessity for a study of heat transfer in the oceans and exchange of energy between the atmosphere and the oceans. The problems of exchange of energy between the oceans and the continents also merit special attention.

3.3. **Radiation budget of the underlying surface.** The limitation of the network of ground actinometric observations and inadequate reliability of the method of computing the radiation budget, determine the importance of the solution of problems related to the simulation of the radiation budget of the underlying surface and its components from the data of satellite measurements of outgoing radiation. The search undertaken from time to time, for a correlation between the outgoing radiation and components of the radiation budget of the underlying surface or the atmosphere, have indicated positive results [24, 44, 80, 85, 127–129, 151, 155, 183]. For example an adequate correlation between outgoing long-wave radiation and effective radiation of the surface, as well as the long-
83 wave budget of the atmosphere was established [24, 113]. However, the problem of establishing the albedo of the underlying surface and short-wave radiation absorbed by the atmosphere appeared to be more difficult. This is precisely why maximum attention during recent years has been given to the solution of such problems.

The total radiation at the level of the underlying surface is determined by assuming the albedo of the underlying surface-atmosphere system, solar radiation absorbed by the atmosphere and albedo of the underlying surface. Out of these, the first parameter is directly measured from satellites, whereas the second can be calculated by taking into account the data on the moisture content of the atmosphere. The albedo of the underlying surface can be estimated from the data of satellite measurement of 'minimum' albedo. All this enables us to work out a method of determination of the total radiation from the data of satellite actinometric measurements. Using the images of cloud cover, the influx of direct solar radiation can also be assessed.

If the albedo A_s of the underlying surface-atmosphere system and albedo A of the underlying surface from the data of satellite measurements (A is determined from the values $A_{s,min}$ for a cloud-free sky), and total radiation Q from the data of ground pyranometric measurements are determined, then the use of the equation of radiation budget of the system,

$$1 = A_s + q_a + (1 - A)Q,$$

enables us to find the shortwave radiation q_a absorbed by the atmosphere. Ellis and Von der Haar [85] parameterized the quantity q_a thus obtained in the following form:

$$q_a = c_1 + c_2 (\tau^{\frac{1}{2}} \ln \tau),$$

where

$$\tau = (1 - n_c) [0.86u^* + 0.14 (1.66\ u)] + n_c (1.66\ u).$$

Here $n_c = (A_s - A_{s,min})/(0.5 - A_{s,min})$ is the effective quantity of clouds, the albedo of which is assumed as 0.5; u the water-vapor content in the vertical column of the atmosphere; $u^* = u \cos z$ (z is the mean-daily zenith angle of the sun); $c_1 = 1.133$ and $c_2 = 0.033$.

The application of parameterization, under consideration, for the computation of q_a offers the possibility of calculation of the total radiation Q at any point of the globe. In [85], an example of application of such a method of revaluation of total radiation from measurement data of brightness of the underlying surface-atmosphere system with the help of a scanning radiometer of the *NOAA-4* satellite for June 1975 has been discussed. The solar constant is equal to 1360 W/m³ and the angular

84 distribution of brightness for the determination of mean daily albedo has been taken into account (the coefficents of transition from measured brightness to radiation fluxes have been found). The analysis of distribution maps of mean monthly amounts of total radiation over the territory of the United States in June 1975, obtained by the application of the method described here, led to the conclusion that the accuracy of revaluating the mean over a month of the daily amount of the total radiation amounts to about 10%. Use of data received from geostationary satellites would increase the accuracy of simulation by taking into account the daily course.

The combination of satellite and ground actinometric measurements makes it possible to determine the shortwave radiation absorbed by the atmosphere. Major [127] carried out calculations of solar radiation absorbed by the atmosphere at different points in Europe, Goose Bay (Canada) and the Azores. The calculations are based on the data of ground measurements of total radiation, albedo of the underlying surface and satellite measurements (*Nimbus-3*) of albedo of the underlying surface-atmosphere system for 8 days in April, 15 days in July, 12 days in October 1969 and 7 days in January 1970. Through the calculations he showed that the coefficient of correlation between the radiation absorbed by the atmosphere and the overall content of water vapor in the atmospheric layer is 0.29. A considerably higher correlation (0.64) exists with albedo A_s of the system:

$$q_a = 32 + 0.21\,A_s - 0.0053\,A_s^2.$$

The absorbed radiation attains a maximum at $A_s = 20\%$ (being 34%), and then decreases with an increase in albedo of the system. If the relationship between absorbed radiation and thickness of the atmospheric mass is also considered, the correlation coefficient rises to 0.81.

The processing of data from satellite measurements of outgoing shortwave radiation collected over 17 seasons (1962–1970) enabled us to draw a world map of the distribution of mean annual values of the albedo of the underlying earth's surface-atmosphere system [183]. In processing this data, results of measurements obtained with the help of wide-angled sensors permitting only the determination of albedo averaged over wide areas, were mainly used. In this case the highly modulated (by spatial averaging) albedo of polar regions was 55–60%.

With the aim of obtaining more detailed information on the geographical distribution of the albedo of polar regions in both the hemispheres, Von der Haar [183] carried out the processing of data of measurements with the help of a five-channel (0.3–3 μm channels) scanning radiometer of medium resolving power (about 50 km above the level of the earth's surface measured vertically downward from the satellite) installed on the *Nimbus-3* satellite. In this case, keeping in view the estimate of albedo of the underlying surface of snow and ice, he studied only the measurement data of the 'minimum' albedo of the system obtained in the absence of clouds or by excluding the effect of cloudiness. The method of processing of measurement data on the intensity of outgoing shortwave radiation, with the aim of determination of the mean daily albedo, is based on assumptions regarding the isotropic nature of the underlying surface and consideration of the daily course of albedo depending on the zenith distance of the sun.

An analysis of maps of geographical distribution of the minimum albedo of polar regions, obtained from data for May 1–15, July 16–31, and October 3–17, 1969 as well as for January 21 and February 3, 1970, shows that the albedo of snow cover above the Antarctic and Greenland often exceeds 60%. The circumpolar isolines of the albedo distinctly reveal the southern boundary of snow cover and pack ice.

Rockwood and Cox [155] considered another possibility of simulating the albedo of the underlying surface in the example of data relating to Africa. Northwestern Africa appears to be the best natural polygon for studying the effect of the nnderlying surface albedo on the general circulation of the atmosphere. Here tropical forests with annual precipitation values reaching up to 200 cm are adjacent to the Sahara desert where the annual rains are less than 10 cm. Between these two zones with extreme characteristics, lies the Sahelian region which is regularly subjected to destructive influences of desert from the north. In this territory the most varied types of underlying surfaces are found: from glittering white sand to various species of vegetation including forests and swamps.

With the aim of studying the relationships concerning space-time variations of albedo, Rockwood and Cox [155] worked out a method of remote sensing of the underlying surface albedo from the data of satellite measurements of outgoing shortwave radiation. The method is based on the application of data obtained by simultaneous measurements of brightness of the underlying surface-atmosphere system from a geostationary meteorological satellite *GMS-1* and aircraft measurements of albedo during the period of GATE tests. The pyranometric (wave band 0.285–2.8 μm) measurements of surface albedo have been taken during flights of *Saberliner* aircraft at an altitude of 150 m [sic]. The type of surface was recorded on cinematic films.

Statistical analysis of the data of aircraft measurements of albedo A_s and satellite measurements of brightness B of the system obtained at 12:00 hours on September 4, 1974 led to the following empirical relationship:

$$A_s = c_0 + c_1 B + c_2 B^2,$$

where

$$c_0 = -1.82454322 \times 10^{-2}, \quad c_1 = 6.722495 \times 10^{-4}$$

and

$$c_2 = 1.70706 \times 10^{-5}.$$

The method under consideration was used for the analysis of variability of the underlying surface albedo in northwestern Africa by comparing the data for July 2, August 10 and September 20, 1974. This comparison revealed high space-time variations of albedo. For instance, at 15°N the albedo dropped from 25 to 19%, i.e. by 25% within 80 days.

Corresponding relative changes at 14° and 17°N latitudes may be as much as 15 and 5%. However, north of 18°N, the relative variation of albedo did not exceed 5%.

The development of a system for classification of the underlying surfaces (on the species of vegetation and extent of its cover) enabled the study of variations of physical properties of the surfaces responsible for variations in albedo. Thus, for example, a surface of class No. 7, practically clear of vegetation, has values of albedo exceeding 42%. Some portions of the Sahara desert, where the albedo has increased to 50%, correspond to this class. For Class No. 3, which is predominant at 15°N in July (combination of vegetation and light-colored soil), an albedo in the range of 21–26% is characteristic. With the beginning of the moist season this class is transformed into class No. 2 (mainly dark-colored vegetation and about 30% bare soil surface), to which still lower values of albedo (16–21%) correspond.

During the period of GATE tests, major changes in the extent of underlying surface sections of different classes took place. Thus, for example, if on July 2 the surface of class No. 6 (albedo 0.36–0.42) covered 24% of the territory under consideration, by September 20 the relative area of this surface dropped to 13%. On the other hand, the area of class No. 3 surface (albedo 0.21–0.26) increased from 4 to 20%. The area of sections with very low and high albedo indicated practically no change. Apparently, the considered variations of surface albedo are related with fluctuations of moisture content (which serves as an origin of precipitation) in the atmosphere. This relationship, however, is still not fully understood.

According to Charney, the increase in surface albedo from 0.14 to 0.35, while moving toward the north from 18°N, causes a 40% decrease in rainfall. The results of the foregoing observations confirm that the zone of the Sahara north of 18°N is characterized by relatively small variations in albedo. Since these small variations cover a vast territory, they may, therefore, be significant. On the contrary, large scale variations are more typical south of 18°N.

87   Raschke and Preuss [151] proposed a new method of restoration of the shortwave budget of the underlying surface and its components from the satellite data of measurements of outgoing shortwave radiation (in the example of results of measurements from *Nimbus-3*) and model calculations. The equation for the relative radiation absorbed by the atmosphere

$$q_a = (1 - A_s) - (1 - A) Q,$$

where Q is the total radiation expressed in fractions of extraterrestrial insolation, and serves as the primary relationship for solution of the problems of revaluation.

Insofar as satellite measurements enable us to determine only A_s (in similar investigations carried out earlier it was assumed that A (or Q) should be determined and Q (or A) revaluated) from data of independent ground measurements; the fragmentary nature of these is responsible for an insufficient reliability of such an approach. The proposed method is based on the introduction of the so-called 'effective albedo' factor:

$$A_c = \frac{A_s - A_{s,\,min}}{A_{s,\,max} - A_{s,\,min}},$$

where $A_{s,\,min}$ and $A_{s,\,max}$ are the minimum (measured in a cloudless sky) and maximum (calculated for conditions of continuous cloudiness) albedos of the system. Insofar as model calculations enable us to determine, for various conditions (oceans, deserts, vegetation cover, different contents of water vapor, aerosol, etc.), not only $A_{s,\,max}$, but also the total radiation Q_0 in dense cloudiness, we have $Q = (1 - A_c)\,Q_c + A_c Q_0$, where Q_c is the total radiation in a clear sky. In this case the establishment of reliability of the computed $A_{s,\,max}$ and Q_0 from the data of actinometric observations is necessary.

Application of the described method has been considered for processing the global data of the *Nimbus-3* scanning radiometer (spectral line 0.2–4.8 μm) for the periods, April 15–August 15, 1969, October 3–17, 1969 and January 21–February 3, 1970. The values of A_s and $A_{s,\,max}$ have been obtained for a spatial resolution of 250×250 km², which is always related with some 'pollution' due to cloudiness and indicates the requirement of a higher resolving power (in the future).

Raschke and Preuss [151] plotted global maps of $A_{s,\,min}$ and A, as well as maps of the total radiation and shortwave budget for the period June 1–30, 1969. On the basis of this data the values of minimum albedo, obtained earlier, above oceans were found to be less than 10%, which are significantly underestimated as compared with the calculated. This can 88 be explained by failure to take into account the daily course of albedo while processing satellite data. The calculations of $A_{s,\,max}$ and Q_0 were carried out with the use of an approximate two-stream method for the model of cloud cover at an optical thickness of 60 for the 0.55 μm wavelength (such type of cloudiness, found at altitudes of 5–6 km, corresponds to the mean daily values of $A_{s,\,max} = 0.68$ at all latitudes).

The revaluated measurements of surface albedo appeared to be overrated for many regions of the globe on account of not eliminating 'pollution' as a result of cloudiness and this gives rise to a significant discrepancy when compared with the global maps of the albedo of the earth's surface drawn earlier. Accordingly, averaged over the zones, the values of total radiation appeared to be overrated by 10–20% and the shortwave budget by more than 10%.

A comparison of maps, drawn on the basis of satellite data of solar radiation absorbed by the atmosphere with available computed data reveals a slight overestimation of computed values. A similar overestimation was noted by Cox earlier (1978) on the basis of analysis of reliable data of aircraft and satellite measurements.

The brightness fields of natural formations, calibrated in absolute units and obtained with the help of multichannel scanning radiometers installed on airplanes and satellites provide fresh prospects for detailed mapping of the radiation budget of the underlying surface and its components. Pease and Nickols [146] discussed the example of automated (computerized) mapping on the basis of data of aircraft measurements for the Baltimore city area (U.S.) at an altitude of 1526 m with the help of a multichannel scanning radiometer on May 11, 1972, at about 13:45 hours local time. The application of measurement data in three spectral intervals (0.62–0.70, 1.0–1.4 and 9.8–11.7μm) helped in drawing maps of natural radiation of the underlying surface in the conditions of the city albedo, absorbed shortwave and long-wave radiation as well as the radiation budget.

The first of these maps was drawn on the basis of data of measurements of radiational temperature for the 9.8–11.7μm channel (emissivity of the underlying surface has been assumed equal to 0.95). The albedo has been found from the data for two shortwave channels. The values of backscattering were taken from the data of ground measurements. The independent ground control (calibrating) measurements of components of the radiation budget which, in particular, enable us to take into account the effect of thickness of the intermediate layer of the atmosphere, were of great importance in drawing the maps. Hence the significance of further improvement in such methods, with the application of data of satellite multispectral measurements, is of great interest.

Chapter II

The Gaseous Composition of the Atmosphere and Radiant Heat Influx

89   The problem of radiant heat influx has been discussed in detail in several monographs [2, 10, 18, 21, 22, 32, 33, 108, 109, 134]. It is understood that the major factors determining the transfer of radiation are as follows (in order of importance): 1) distribution of cloudiness; 2, 3, 4) vertical profiles of temperature, water vapor and ozone, 5) aerosol, 6) spectral properties of the underlying surface, 7) concentration of carbon dioxide, 8) altitude or pressure at the lower boundary layer of the atmosphere and 9) other small gaseous components. In a number of cases it appears that a mutual interaction among these factors cannot be ignored.

The selection of an adequate scheme of parameterization of radiant heat influx in numerical modeling of the general circulation of the atmosphere or of climate is resolved while keeping in view the required accuracy of such a scheme and accordingly, the possibility of simplification. From this standpoint a classification of computational methods, reproduced in Table II.1, proposed by Rodgers [146], is possible. Here 'climatology' means the use of climatic mean values, and a gap represents an absence of consideration of the respective factors. Practically all the alternatives of the method mentioned in Table II.1, were used in numerical modeling of the general circulation (even for direct line-by-line calculation).

Let us now examine some new results with reference to the problems of current changes in climate and related with determination of gaseous components of the atmosphere and radiant heat influx on account of molecular absorption.

Although it has often been asserted that the quantitative characteristics of the field of thermal radiation can be evaluated with an accuracy fixed before hand, they should not be considered acceptable for a

90

Table II.1. Methods of computation of radiative transfer (in order of increasing complexity and decreasing speed of computations)

Shortwave radiation		Long-wave radiation	Remarks
Scattering	absorption		
Climatology	Climatology	Climatology	Without feedback links
Empirical relationships based on observational data or calculations	Empirical relationships based on observational data or calculations	Newton's law of cooling	Feedback links with respect to temperature
	Total absorption by each gas, taken separately into account	Various empirical relationships	Feedback links with respect to temperature
Approximation of mean free path of photons	As in case of long-wave radiation	Emittance (similar to radiation nomograms)	Number of feedback links may exist
Method of doubling or integration with respect to frequency, etc.	Line-by-line computations	Spectral calculations for many intervals. Approximation of Curtis-Godson model of poles	
		Line-by-line computations	Tedious calculations

number of reasons: 1) there exist considerable uncertainties in spectroscopic data of high resolution in absorption spectra (this especially concerns the half-width and contours of lines), 2) all possible factors determining the transfer of thermal radiation have not been sufficiently studied, 3) till now a sufficiently correct comparison of results of measurements and calculations has not been achieved. On the other hand toward the beginning of the sixties three important circumstances enabled us to solve the problem of calculation of fluxes and influxes of long-wave radiation in a new way: 1) fast progress in the investigations and parameters of high resolution absorption spectroscopy for various atmospheric trace species, 2) access to high speed computers, 3) possibility of having comprehensive observational data for testing computational methods.

Ellingson and Gille [72] undertook the modification of the method of calculation of fluxes and influxes of long-wave radiation proposed by Rodgers and Walshaw in 1966, the main aim of which was to develop a standard method on the basis of more detailed calculations (from the viewpoint of spectral resolution and consideration of the optics and active gaseous components of the atmosphere), based on the application of currently available information on the quantitative characteristics of absorption. The computation was carried out with the consideration of the rotational and 6.3 μm lines of water vapor, the 15 μm line of carbon dioxide, the 14 μm and 9.6 μm lines of ozone, the 7.66 μm line of methane, and the 7.78 μm line of nitrous oxide. Absorption in the lines of water vapor, carbon dioxide and ozone is described by the statistical model of Gudi and in the bandwidths of methane and nitrogen monoxide by empirical exponential transmission functions. The exponential function approximates the continuum of absorption by water vapor at 400–1200 cm^{-1}.

In order to take into account the effective vertical heterogeneity of the atmosphere the approximation of Curtis-Hudson is used. The spectrum in the range of wave numbers 0–2814 cm^{-1} is divided in one hundred intervals of unequal width: from 2–3 and 5 cm^{-1} at the centers of Q-branches to 40 cm^{-1}. The shape of the Lorentz spectral lines is considered. The thickness of the atmosphere is considered from 0–60 km divided into several layers of varying thicknesses (the range of altitudes is indicated in brackets): 0.3 km (0–12 km); 0.5 km (12–30 km); 1.0 km (30–48 km) and 2 km (48–60 km). The constant mixing ratios of carbon dioxide, methane and nitrous oxide are taken as 320, 1.75 and 0.28 ppm respectively. The vertical profile of ozone is taken as the climatological profile for 15°N. The effect of partial cloudiness has been considered by the method of Manabe and Strickler. Instead of integration with respect to angles in the calculation of fluxes, the diffusion coefficient equal to 1.667 was used.

The comparison (with the help of an infrared interference spectrometer) of spectral distributions of outgoing radiation in a clear sky, measured from *Nimbus-3*, with computed values revealed, as a rule, agreement within limits of 10%. In [72], the possible reasons for discrepancies between the results of calculations and observations are discussed 92 in detail. If those regions of the spectra are considered in which the outgoing radiation was not measured, it can be concluded that the accuracy of calculations of total outgoing radiation in a clear sky is not less than 3%.

It is important to note that the occurrence of systematic errors of radiosonde measurements of relative humidity (5–10%) and temperature (about 1 °C) leads to errors in calculations of outgoing radiation of up to 1.0 and 1.7%, respectively. The random errors of aerological data cause errors of outgoing radiation of up to 0.4%. Thus, in addition to the lack of reliability of initial data on the temperatures in high resolution absorption data, the errors in aerological data alone introduce errors in calculations of the total outgoing radiation of up to 3%.

If methane and nitrous oxide are ignored, this changes the outgoing radiation in conditions of a tropical atmosphere by slightly more than 1%. In this case, the variation in radiative changes of temperature is only few hundredths of a degree in a day. On comparison of calculated values with the data of an actinometric radiosonde (ARS), considerable discrepancies were observed. Thus, the calculated downward fluxes of radiation at all altitudes are underrated as compared with observed values. The still unsolved problems of accuracy of an ARS make it difficult to explain the reasons for these discrepancies. Further modification of computational methods is possible only with the data of high-precision measurements of spectral fluxes of radiation and sufficient data of aerological soundings.

In [17, 19, 22], an approximate method for calculation of the intensity of thermal radiation of the atmosphere, meant for use on a computer, is proposed. The method is universal in the sense that it is suitable not only for the conditions of the earth's atmosphere, but also for other planets.

1. PARAMETERIZATION OF RADIATION PROCESSES IN THE GENERAL CIRCULATION MODEL OF THE ATMOSPHERE

Rapid progress in the field of numerical modeling of the general circulation of the atmosphere (GCM) and climate has stimulated the development of most varied 'radiation blocks'. Here we shall limit ourselves to individual illustrations with regard to this aspect, keeping in mind that it has been discussed in detail, for example, by Gudi [2], Kondrat'ev [10, 108, 110], and Feigel'son [32, 33]. The first question that arises in con-

nection with the development of the 'radiation block' of the GCM lies
93 in the degree of sensitivity of the parameters in the considered model
to radiation. An answer to this requires numerical modeling for analysis
of the sensitivity of models to various parameters (including radiational).
It should be noted that such modeling is still in its initial phase. Only for
one- and two-dimensional models of the general circulation of the atmosphere have preliminary, though relatively comprehensive results, been
obtained.

Reck [143, 144] discussed the results of calculations of sensitivity of
the vertical profile of temperature in the framework of a one-dimensional
model of radiative-convective equilibrium to the variations of certain input parameters of the model. The results related to the troposphere indicate that:

1) except for $\partial T/\partial p$ (p—pressure) and $\partial T/\partial G$ (G—vertical temperature
gradient), the temperature T_s of the underlying surface is most sensitive
to the input parameters;

2) the effect of all parameters on temperature, except for atmospheric
pressure, vertical temperature gradient and the Rayleigh scattering,
attenuates linearly with altitude, attaining approximately 80% of the
ground value at the level of 336 mb (8.3 km);

3) the effect of the Rayleigh scattering on temperature decreases with
altitude, two times faster, becoming 64% at the level of 336 mb;

4) the reaction of temperature to changes in pressure increases with
altitude; and

5) with the consideration of aerosol, the sensitivity of temperature to
such parameters as relative humidity, concentration of carbon dioxide,
surface albedo and the solar constant, decreases.

Calculations for the stratosphere show that:

1) the reaction of temperature to changes in 'tropospheric' parameters
(for example, relative humidity at the surface, Rayleigh scattering, surface albedo and the aerosol coefficient of attenuation) declines in transition from the lower to upper stratosphere;

2) the effect of 'atmospheric' or stratospheric parameters (for example,
concentration of carbon dioxide, pressure, solar constant, concentration
of water vapor in the stratosphere) on temperature increases with altitude
in the stratosphere parallel to an increase of temperature with altitude; and

3) the sensitivity of temperature to the concentration of ozone attains
a maximum, approximately at the level of 74 mb.

Table II.2 characterizes the sensitivity of T_s to various physical parameters for a change of each of the parameters by 1%, while the value of
$\partial T/\partial S_0$ (S_0—solar constant) has been taken as unity. Here the probable
errors in the measurement of parameters and the uncertainty in T_s, caused
by these errors are also well presented.

Table II.2. Sensitivity of temperature of underlying surface to variations in certain parameters by 1% and errors of input parameters

Parameters	Value	Relative sensitivity	Variation of parameter causing change of T_s by 0.1 K	Probable error of measurements	Uncertainty of T_s caused by error of measurements
Solar constant, W/m²	1365	1.0	0.002	0.01	0.5
Zenith, angle of sun, deg	59.20	0.98	0.04	0.04	0.1
Duration of day, min	720.00	0.90	0.7	0.7	0.1
Heat capacity of dry land surface, J/K gm	1.0048	0.68	0.001	0.002	0.2
Surface albedo, %	14.0	0.11	0.1	0.5	0.5
Relative humidity at the surface, %	74.0	0.11	1.1	5.0	0.5
Vertical gradient of critical temperature, K/km	−6.50	0.10	0.02	0.03	0.1
Atmospheric pressure at ground level, mb	1000.0	0.077	11.0	4.0	0.03
Coefficient of Rayleigh scattering, %	7.0	0.069	0.09	0.1 (?)	0.1
CO_2 concentration, ppm	496.0	0.018	23.0	2.0	0.01
Aerosol coefficient of attenuation, km⁻¹	0.100	0.0055	0.016	0.002	0.01
O_3 concentration, %	100	0.0028	31.0	1.5	0.005
Pressure at 'dry boundary', mb	20	0.0016	11.0	10.0	0.1

94   The results thus obtained, indicate that the temperature of the underlying surface is more sensitive (by an order of magnitude) to changes in the underlying surface humidity and relative humidity of the air, near the surface, than to variations of components of the atmosphere such as carbon dioxide, ozone and aerosols. Although changes in the zenith angle of the sun, the duration of the day, thermal conductivity, critical vertical temperature gradient, atmospheric pressure and Rayleigh scattering are considerable, the fact that all these parameters are determined with a sufficiently high accuracy, is also significant.

    The results examined here enable formulation of requirements toward the accuracy of determination of various parameters from the viewpoint of calculation of T_s with an error not exceeding 0.1 K. As can be seen, 95 increasing the accuracy of measurements of the surface albedo, solar constant and relative humidity is of considerable significance.

    Chow and Curran [55], on the basis of a modified version of the zonal model of mean annual climate developed by Chring and Adler [131], evaluated the effect of selection of various methods of parameterization of evaporation on the sensitivity of climate to variations in the solar constant. A new nonlinear parameterization (model A), in particular, is considered in which the Bowen ratio is expressed as a function of the vertical gradient of humidity and temperature near the earth's surface.

The dynamic model is based on the equations of a two-level quasi-geostrophic potential vortex, which takes into account the radiant heat influx at the cost of shortwave and long-wave radiation, convection, phase transformations of water and transfer of heat to the oceans.

The analysis of computed components of the thermal budget showed that:

1) results for a linear model B of parameterization and model A are extremely close, except for the distribution of energy between the fluxes of heat on account of the turbulence and phase transformation of water;

2) calculated values of absorbed atmospheric solar radiation are underestimated as a consequence of ignoring the absorption, which is caused by carbon dioxide, oxygen, clouds and aerosols;

3) radiation absorbed by the earth's surface, which determines the turbulent heat-exchange and effective radiation of the earth's surface is overestimated; and

4) the overestimated effective radiation calls for a decrease in radiational cooling of the atmosphere.

The analysis of calculations of change in temperature of the surface at the 500 mb level in the case of a decrease of the solar constant by 1%, reveals maximum variations at higher latitudes which is due to the effect of an albedo feedback link. Model A leads to results similar to those obtained earlier by Manabe and Wetherald [121], albeit somewhat underestimated.

The data presented in Table II.3 characterizes the sensitivity of the components of the thermal budget of the underlying surface to a decrease in the solar constant by 1% for models A and B. Here ASR represents solar radiation absorbed by the underlying surface, ER the effective surface radiation, THF the turbulent heat flux from the underlying surface to the atmosphere, UHF the upward heat flux on account of evaporation 96 from the surface, and GCMA the general circulation model of the atmos-

Table II.3. Variation of components of thermal budget of underlying surface (%) for decrease in solar constant by 1% for models A and B

Component	Model A	Model B	GCMA
ASR	−1.2 (176.8)	−1.3 (176.4)	−1.2 (166.0)
ER	+0.1 (67.1)	−0.4 (67.8)	+1.3 (63.5)
THF	+2.9 (26.4)	−1.0 (37.1)	+2.2 (27.2)
UHF	−3.6 (83.3)	−2.3 (71.5)	−4.3 (75.3)

phere. The average values of components of the thermal budget (W/m^2) for the northern hemisphere are indicated in brackets.

As can be seen, the models A and B of parameterization lead to differences not only in values, but (sometimes) also in signs of changes of components of the thermal budget. Only the values of effective radiation, calculated with the application of model A considerably differ from the data for GCMA (the main reason for this divergence is the assumption in regard to the unchangeability of relative humidity in model A). In all, the obtained results demonstrate that adequate parameterization of evaporation plays an important role from the standpoint of estimation of the sensitivity of climate to variations of the solar constant.

Schwarzkopf and Wetherald [152] described the scheme of parameterization of radiation processes developed at the Laboratory for Geophysical Hydrodynamics at Princeton University for a numerical three-dimensional model of general circulation of the atmosphere, relating to only one-third of the hemisphere (sectorial GCMA). The aforementioned scheme is supposed to replace the method of Manabe-Strickler used earlier and is based on the application of several versions of the Lacis and Hansen (LH-74) method [112].

The main objective of the application of the new method of parameterization of radiation processes is to have a more accurate account of absorption by water vapor and ozone, Rayleigh scattering and multiple reflection of solar radiation by clouds in conditions of partial cloudiness. In the case of more than one layer of cloudiness, the albedo of low-lying clouds is referred to as the albedo of the underlying surface.

Schwarzkopf and Wetherald [152] compared the old and new methods of parameterization for nine latitudinal belts. Method LH-74 leads to increased values of the mean global effective flux of shortwave radiation at the level of the upper boundary of the atmosphere by approximately 97 10 W/m^2. As a consequence, the global albedo decreases from 32 to 30%. At the same time, an increase in the effective shortwave flux near the earth's surface at all latitudes is observed and is about 15 W/m^2. The radiational warming of individual layers of the atmosphere did not undergo significant changes.

The comparative numerical modeling for the general circulation of the atmosphere with the application of old and new radiational parameterization with an unchanged solar constant revealed an increase in average zonal temperature of air near the earth's surface, at all latitudes, by more than 9 K in the case of application of the LH-74 method, which is mainly due to the multiple reflection of solar radiation by clouds. Simultaneously, there is a fall in the meridional gradient of surface temperature by 8 K and a drop in the temperature of the stratosphere by 2 K (the latter is caused by a decrease in absorption of solar radiation by ozone). Application of the LH-74 method reduces by 4%, the sensitivity of the temperature field to the increase in the solar constant in the troposphere of temperate latitudes from 3.2 to 2.9 K (averaged values).

Wu et al. [182] carried out numerical experiments with the aim of estimating the sensitivity of the GCMA to vertical profiles of radiant heat influx (RHI) at the cost of long-wave radiation calculated by the method of Hogan or Wu and Kaplan. The experiments were based on application of a 9-level model of general circulation of the atmosphere, developed by a team at the Goddard Space Flight Center, NASA.

Special features of the Wu-Kaplan method [181] appear to be the description of transmission in the zones of water vapor with the help of a statistical model of absorption bands, the application of the Curtis-Hudson approximation in order to take into account the inhomogeneity of the atmosphere and the effect of absorption by dimer in the spectral window of 8–13 μm. The transmission by carbon dioxide has been calculated beforehand by direct integration (line-by-line). The absorption by ozone, considered only during summer, has also been calculated in advance. Such parameterization corresponds with the results of accurate calculations (line-by-line) and provides agreement within the limits of errors of measurement between calculated values and those measured with the help of actinometric radiosondes giving vertical profiles of fluxes of long-wave radiation.

The consideration of dimer absorption is manifested as a significant increase in radiational cooling in the lower troposphere at lower latitudes. The dimer absorption promotes destabilization of the lower troposphere at lower and temperate latitudes. The additional latent heat released as a result of this destabilization can be a new source of energy for a dynamic process at lower and temperate latitudes.

The numerical modeling of general circulation of the atmosphere was carried out for the given initial state on the basis of actual data from December 20, 1972, and June 18, 1972 onward, for periods of 70 and 72 days, respectively, with averaging of investigated parameters over the previous 30 days. A systematic discrepancy in the computed results of mean zonal temperature, location and shape of the maximum of zonal westerly winds, intensity of the Hadley and Ferrel cells of circulation, distribution of clouds and precipitation and also the distribution of mass were observed.

The comparison of calculated data with the results of absorption for assessment of the adequacy of each method and effect of the consideration of radiant heat influx on the results of numerical modeling of the general circulation of the atmosphere were also carried out. It has been found, for example, that application of the Wu-Kaplan method enables us to explain characteristic peculiarities of the GCA—such as an Icelandic low and monsoonal rains. This, however, cannot be done by the Hogan method. In some cases the application of described methods of calculating radiant heat influx leads to differences, which are less than the discrepancies between calculated and observed data. However, in all such cases the Wu-Kaplan method offers better results, which permits us to consider this method of parameterization of radiant heat influx as more reliable.

A discussion of the results convincingly demonstrates that the problems of adequate selection of radiation blocks in numerical modeling of the general circulation of the atmosphere and climate has significant importance. Let us now briefly discuss some schemes for computing radiant heat influx due to long-wave radiation.

The method of 'exchange between the layers' proposed earlier for the calculation of spectral radiant heat influx due to long-wave radiation was used by Joseph and Burtsztyn [105]. The method has been used for calculation of the total (from the spectrum) radiant heat influx with the aim of using such a method in the framework of the numerical model of general circulation of the atmosphere developed by Mintz and Arakava. Joseph and Burtsztyn [105] carried out a detailed comparison of four components of radiant heat influx. These were cooling on account of radiation in space, warming by way of heat exchange with the underlying surface and two components characterizing radiant heat exchange with the layers of atmosphere located above and below the considered layer. The consideration of radiant heat exchange between the layers presents great difficulty, and an approximate parameterization of such an exchange, suitable for specific analytical presentation of vertical profiles of temperature and humidity (lower, subtropical and higher latitudes in the Mintz and Arakava model of general circulation of the atmosphere) is proposed. The application of such parameterization in other models of general circulation of the atmosphere only requires modification of the constants (the assumption in regard to decrease of absolute humidity with altitude, is of prime significance). Another possible method of parameterization consists in assuming a radiant heat exchange between the layers by including it in the heat exchange at the boundaries. However, the first of these methods appears preferable, since it is accurate enough and provides considerable economy in computer time.

In the study of radiational cooling of the atmosphere (RCA) caused by long-wave radiation it is important to distinguish between radiational cooling of the atmosphere determined by the effect of the atmosphere and the perturbation of RCA related with radiation damping of perturbations in relation to some basic state. Traugott [171] considered perturbation of RCA belonging to two categories: independent and dependent on spatial scales. For disturbances of the first category, their velocity can be found by differentiation of the localized values of RCA from temperature and such an approach is justified in a wide range of conditions. Since in this case the scale of perturbation is large as compared to the characteristic scale of the atmosphere, such disturbances have been called macrolevel disturbances.

The disturbances caused by spatial harmonic variations and attenuation with the increase of wavelength of the perturbation (except in the case of very small optical thicknesses for short wavelengths) serve as an example of excitation of RCA of a significant category. Such disturbances have been called microlevel disturbances.

The disturbances of the two aforementioned categories should be considered as a limiting case of excitations of RCA in conditions when the characteristic scale of the disturbance is considerably higher or lower than the internal scale of the atmosphere. Traugott [171] proposed an approximate method that can be used to calculate RCA as a result of disturbances of the temperature field in an inhomogeneous atmosphere. It is also assumed that these disturbances possess arbitrary horizontal and vertical wavelengths. This enables us to describe the transition zone between disturbances of RCA in a homogeneous medium depending on the spatial scales and overall disturbances in radiational cooling of the atmosphere independently of the scales. The latter prevail in the case where the wavelength appears to be shortest in the vertical direction and exceeds the altitude of the homogeneous atmosphere; to the first category belongs the limiting case of a wavelength which is considerably shorter than the altitude of the homogeneous atmosphere.

The use of computational methods relating to any of the limiting cases beyond the regions of their applicability leads to an underestimation of the earth's radiation budget. It has been shown that the approximate method under consideration, based on the Milne-Eddington approximation and averaged with respect to the frequency of the coefficient of absorption leads to results which agree well with the results of more accurate calculations even with a deviation from the localized thermodynamic equilibrium.

The calculations of absorption of radiation by water vapor, carbon dioxide and ozone with the use of a statistical model of the absorption band showed that, within the limits of every absorption band, there is an exponential change in the generalized coefficient of absorption l depending on the wave number. This enables us to approximate the transmission function for an arbitrary spectral interval by the exponential of integral $E_1 (\sqrt{lm})$, where m is the absorbent mass. Such an approximation has been used by Kuo [111] in the calculations of vertical profiles of various components of total radiant heat influx for stratification, corresponding to a standard model of the atmosphere.

94

The calculations led to the conclusion that the vertical profiles of radiational cooling caused by $H_2O + CO_2 + O_3$, $H_2O + CO_2$, $H_2O + O_3$, and $H_2O + O_3 +$ double CO_2 content, are similar and characterized by the occurrence of a maximum of about 1.3 °C/day near the earth's surface, a minimum at an altitude of 2 km, a secondary maximum in the vicinity of 7.5 km and a minimum near the tropopause. In the stratosphere there is a monotonic increase (at altitudes of up to 40 km) to total radiational cooling with increasing altitude, whereas with the consideration of H_2O $+ CO_2$ there appears a small minimum at the altitude of 32 km. The radiational cooling due only to water vapor is maximum (1.6 °C/day) near the earth's surface, drops to 1.16 °C/day at an altitude of 4 km, and remains practically unchanged in the 4–3 [sic] km layer, after which it drops to its minimum value at the level of the tropopause. In the stratosphere the maximum is noted at an altitude of 23.5 km.

The cooling in the troposphere caused by carbon dioxide is approximately six times less than the same caused by water vapor, but becomes predominant above 28 km. Ozone causes a radiational warming in the 6–31 km layer. The doubling of carbon-dioxide content leads to a decrease of cooling in the lower troposphere by less than 0.03 °C/day. Though the main component of the radiant heat influx appears to provide cooling in space, the radiant heat exchange with the underlying surface and other layers of the atmosphere as well, provides a significant contribution.

As observed by Coakley and Briegleb [60], the calculations of longwave radiation fluxes and radiant heat influx with the use of data on the emittance of the atmosphere enable us to realize a considerable economy in computer time, but are not proved to be sufficiently accurate. However, errors attributed to the method of calculation of emissivity, occur quite often, due to assumptions made for the calculation of emissivity on the basis of data on absorption (approximated by the models of absorption bands). Seldom is it assumed, for example, that the emissivity depends only on the mean weighted (with the consideration of pressure) amounts of matter absorbing the radiation (strong line approximation), whereas the optical path length does not depend on temperature. Coakley and Briegleb [60] showed that the last assumption appears to be a source of serious error in the calculations of radiation fluxes and heat influx. They proposed a method of consideration of the temperature relationship, which enables us to increase the accuracy of calculation of effective radiation near the boundaries by 3%, and of radiational cooling by 0.04 K/day. The climatological vertical profiles of radiational cooling for tropical, temperate and polar latitudes in a cloudless sky have been obtained.

The detailed account of radiant heat influx on account of long-wave radiation in the numerical modeling of atmospheric processes is related with the necessity of using up a longer computer time as compared with that required for consideration of other processes. Hence the urgent necessity of developing such methods of parameterization of radiant heat influx that can be carried out economically, but at the same time provide rapid and reliable computation of radiation fluxes. Chow and Arking [54] carried out a generalization of the results obtained earlier in this direction with the aim of covering the entire spectrum of thermal radiation in a separate study of the water-vapor band and the 15 μm line of carbon dioxide.

For water vapor, the line absorption as well as continuous absorption of the e type is kept in view. In the first case the use of a line edge approximation enables us to introduce the transmission function in the form of an exponent which is calculated beforehand using the line-by-line method. Subsequently, calculation of the long-wave radiation fluxes can be carried out rapidly and accurately.

The comparison of vertical profiles of radiational cooling, thus obtained, for tropical and sub-tropical atmospheres during winter with the results of accurate calculation indicated a maximum error of less than 0.2 °C/day. A similar method is also proposed for the 15 μm wavelength of carbon dioxide, except that in this case consideration of the temperature relationship of absorption, within the framework of approximation of linear disturbances, is essential. The overlapping of wavelengths is taken into account by using the rule of multiplication of transmission functions. The radiant heat influx in the region of the 15 μm wavelength manifests itself mainly as radiational cooling in the upper stratosphere and above. The maximum error in calculation of radiant heat influx does not exceed 0.3 °C/day.

Fels [81, 82], worked out several algorithms providing quick and accurate calculations of radiational cooling in the stratosphere and the mesosphere as a result a radiant heat exchange in the 15 μm wavelength of carbon dioxide, which can be used in numerical modeling of atmospheric processes. The uniform mobility of carbon dioxide in the atmosphere allows us to carry out computations of the transmission function beforehand by using an accurate method (line-by-line). This renders the calculations of radiation fluxes and heat influxes almost trival.

The solution of the problem of calculation of transmission functions, becomes complicated due to the necessity of having to take into account their temperature relationship, arising because of the temperature dependence of line intensities. Therefore, Fels [82] proposed consideration of the temperature relationship of transmission in the linear approximation, which reduces the problem to the computation of the transmission function for a standard stratification τ_s and its derivative $\partial\tau/\partial T$. In this case the transmission function is determined as:

$$\tau(p_i, p_j) = \tau_s(p_i, p_j) + \partial\tau(p_i, p_j)/\partial T \, \Delta\overline{T},$$

where $\Delta\overline{T}$ is the effective change of temperature (in relation to the standard profile), and p_i and p_j are the atmospheric pressures determining the boundaries of the atmospheric layer under consideration.

Fels [82] carried out accurate (line-by-line) calculations of $\tau_s(p_i, p_j)$ and $\partial\tau(p_i, p_j)/\partial T$ taking into account 1) changes in the intensity of lines along the path of the ray, 2) the Foight profile above the 100 mb level, 3) approximately 4000 lines belonging to 19 bands of the 15 μm complex in the frequency range of 500–850 cm^{-1}, and 4) integration with respect to the angles instead of the use of the coefficient of diffusion.

The carbon-dioxide mixing ratio has been assumed as 330 ppm (by volume). The calculations cover about 100 layers in the altitudinal range of 0–85 km. Accuracy of final results depends on the reliability of assigning the parameters for the fine structures in the absorption spectrum. A comparison with the results of calculations carried out by Dickinson showed that the maximum differences observed at an altitude of 50 km attain 0.4 °C/day for a radiational cooling of about 8 °C/day.

Undoubtedly, the main complexity of the problem of the stratosphere arises because of the interdependence of various processes, as examples of which the following may be considered (Callis [49]):

1) interaction between the thermal regime and chemical processes caused by the temperature relationship of the rates of chemical reactions;

2) effect of the underlying surface albedo, lower layers of the atmosphere and clouds on the chemistry of the stratosphere related with scattering and reflection of solar radiation;

3) dependence of meridional winds on the propagation of global waves;

4) link between meridional profiles of temperature and zonal winds;

5) effect of trace elements in the stratosphere on the thermal regime and thermal budget;

6) influence of the stratosphere on climate in the troposphere;

7) effect of variability of ultraviolet solar radiation on the chemistry of the stratosphere and temperature; and

8) dependence of dynamics of the stratosphere on concentration of minor components.

Callis [49] discussed in detail some of the aforementioned examples of the interdependent nature of atmospheric processes. Particularly, the consideration of the temperature relationship of reaction rates, which is often presented in the form proposed by Arrhenius, is also important:

$$k = A \exp(\pm B/T).$$

For large values of B, the sensitivity of k to temperature T appears to be very high. If $B=2500$ (this refers, particularly, to the important reaction $O+O_3\rightarrow 2O_2$), then a change in temperature $\Delta T=25$ K involves a change in the reaction rate by more than 100%. In case of reactions, $NO+O_3\rightarrow NO_2+O_2$ $(B=-1450)$ and $O_2+O+M\rightarrow O_3+M$ $(B=510)$ at $\Delta T=20$ K, the rate of destruction of ozone by the formation of nitrogen oxide increases by 50%, and the formation of ozone decreases by 20% (this data relates to $T=240$ K corresponding to an altitude of 34 km).

Most significant, of course, is to assess the resultant temperature sensitivity of combined reactions. The assessment with the consideration of $O_x\leftrightarrow HO_x\leftrightarrow NO_x$ cycles revealed the existence of a significant negative correlation of ozone concentration and temperature at altitudes above 25 km which attain maximum values at the level of 40 km (here increase 104 in temperature by 3% causes a drop in the ozone concentration by 12%). This is important since it is precisely at these levels that ozone is most sensitive to the action of chlorofluorocarbons.

Through the temperature relationship the effect of variation in the concentration of trace elements of the stratosphere influencing the thermal regime is revealed. Thus for example, doubling of the carbon-dioxide concentration may cause a drop in temperature at an altitude of 50 km by 10 K, which should enhance the ozone concentration (up to a maximum of 16% at a level of 43 km). The increase in total ozone content for a doubling of the carbon-dioxide concentration is about 1%. The growth of ultraviolet solar radiation in the course of an 11-year cycle causes an increase in ozone concentration attaining a maximum value of 17% at an altitude of 40 km. The consideration of the inverse relationship with an increasing temperature reveals a change of the sign of influence on ozone (decrease of concentration) above 42 km.

As indicated by Sundararaman [165], the analysis of one- and two-dimensional models of the atmosphere applied for the assessment of its anthropogenic impact in the stratosphere reveals different sources of errors. Of these the most significant are related with the selection of an adequate system of reactions and data on rates of reactions. There exist, other sources of errors which, particularly, include the following:

1) neglect of changes in extraterrestrial ultraviolet solar radiation;

2) lack of a reliable description of the 'natural' budget of ozone;

3) use of data on vertical profiles of the coefficient of turbulent diffusion obtained on the basis of variations of different traces and this data not being sufficiently exhaustive and reliable;

4) impossibility of accurate determination of leaching of minor components in the troposphere by precipitation;

5) complexity in assessment of the effect of emissions by aircraft of

nitrogen oxides in the vicinity of the tropopause (an insufficiently studied level of the most intensive dynamics;

6) fragmentary nature of observations of trace-element concentrations;

7) lack of a reliable determination of the effect of multiple scattering on the concentration of reacting components, as well as time and spatial averages of photodissociation rates; and

8) incompleteness and insufficient reliability of information on the concentration of such components as carbon dioxide and ClX, at present and in future, in view of the importance of feedback caused by temperature effect on account of carbon dioxide and chemical activity of a small 105 quantity of chlorine. The presence of such feedback links demonstrates the necessity of bearing in mind the complicated nature of these mutually interacting processes.

Various approaches were developed for assessment of the effect of given variations in the ozone content on the climate. The methods varied from the simple one- or two-dimensional models (see, for example, Coakley [61], Schoeberl and Strobel [151]) to complicated three-dimensional models. The calculations carried out earlier with the application of a comparatively simple model showed that the effect of a decrease in ozone content on the temperature of the earth's surface is determined by two contradictory factors:

1) an increase in absorption of solar radiation by the troposphere; and

2) a decrease in nonconvergent thermal radiation in the atmosphere.

It may be assumed that the presence of two mutually compensating factors of comparable values should determine the high sensitivity of calculated surface temperature to detailed peculiarities of the accepted method of computation of radiation fluxes. In this connection Coakley [61] carried out fresh calculations for the effect of a decrease in ozone content on the underlying temperature with the assumption of a uniform decrease in ozone content in the 12–40 km layer or shifting of the height of maximum concentration for an unchanged total ozone content. The results thus obtained, indicate the differences in the estimates (fresh estimates of decrease in surface temperature being about 35% of the preceding values) partially caused by a higher sensitivity of temperature to the ozone content and more considerable changes of the earth's albedo in the earlier model.

The important factor determining the discrepancy of results is the difference in the degree of influence on the ozone content in the troposphere. This difference determines the specific nature of the impact on tropospheric ozone in the models under consideration. The surface temperature appeared to be very sensitive to small variations of ozone concentration in the troposphere. Changes of surface temperature caused by variations in the vertical profiles of ozone are also partially the result of variations in the tropospheric ozone content.

In recent years, the faculty of the Laboratory for Geophysical Hydro-dynamics (GHL) at Princeton University developed the new three-dimensional 40-level model of general circulation of the troposphere-stratosphere-mesosphere system (altitudes 0–80 km) while taking into con-
106 sideration the underlying surface topography. The horizontal latitudinal-longitudinal grid of the model takes into account the Fourier filtration of variables at higher latitudes. A new parameterization of surface transfer taking into account the relationship with the Richardson number and grid-scale effects has been applied. The model contemplates parameterization of thermal damping as a result of radiative-photochemical interaction but avoids calculation of the ozone concentration as a totally dependent variable. The concentration of ozone is given but may 'fluctuate' depending upon local disturbances of the temperature field. The warming due to absorption of solar radiation influences a change in the ozone field causing a damping of temperature disturbances.

Mahlman et al. [120] carried out numerical experiments with the aim of assessing the reaction in the stratosphere due to the considerable changes of ozone content for fixed (on the climatic data) amounts of cloud and temperatures of ocean surfaces. In order to obtain a preliminary assessment, they assumed a low resolution on the horizontal plane (9° latitude × 10° longitude) and a mean annual insolation (in the future, it is proposed to remove these constraints). The calculations have been carried out for an ozone content reduced by two times as compared with the known climatic values.

The control calculations were carried out for an initially undisturbed isothermal atmosphere with a proportion of water vapor (at all points of the atmosphere) equal to 2×10^{-6} g/g). A state of equilibrium was attained after a lapse of 500 days which is assumed as the starting point for calculations with half this ozone concentration while all other conditions remained the same. For these (and control) experiments the integration was carried out over an interval of 160 days with an averaging of the final results over the last 60 days. In spite of the low horizontal resolution, the data for the control experiment qualitatively agrees with mean annual climatic conditions. The higher vertical resolution even provided for elimination of 'chronic' defects of the earlier modification of the GHL models, which involved an underestimation of the temperature of the stratosphere at higher latitudes.

Numerical modeling reveals considerable (about 23 K) global cooling of the upper stratosphere and lower mesosphere. The lower values of cooling of the rest of the atmosphere vary from 1–6 K in the 7–18 km layer

100

to 6–8 K at altitudes of 18–40 km. The other important effect of decrease in ozone content is a drop in the altitude of the stratopause by approximately 6 km, but also an elevation of the tropical tropopauses by 1–2 km. At the same time the mean contrast of temperature between the equator and high latitudes also decreased considerably, by 10 K in the vicinity of the stratopause and approximately by 2 K in the 20–35 km layer.

107   In agreement with the observational data, the meridional temperature gradient at altitudes of 12–20 km has an opposite sign and its value decreases in this case by 2–3 K (cooling at higher latitudes is more severe than in the region of the tropical tropopause). Since the mean zonal winds are almost geostrophic, a decrease in the meridional contrast of temperature exerts a considerable effect on the wind field. The velocity of the maximum zonal western winds in the mesosphere decreases by 10–15 km and in the upper and middle stratosphere by 5 and 1–2 m/s. respectively. For a quantitative assessment of variations of temperature and wind fields in the troposphere further modifications of the model are essential.

2. THE GREENHOUSE EFFECT OF THE ATMOSPHERE*

It is well known that the radiation regime of the atmosphere is determined to a considerable degree by the presence of optically active components in it, such as water vapor, carbon dioxide, ozone and aerosol, whereas one of the principal mechanisms of the effect of radiation factors on climate is related with the so-called greenhouse effect of the atmosphere. Being relatively transparent in regard to solar radiation, to a considerable extent the atmosphere prevents loss of heat on account of emission from the earth's surface into space. Only a back radiation of the cloudless atmosphere compensates to a small degree for thermal radiation from the earth's surface. Precisely for this reason, the study of optical properties of the atmosphere in the window regions and in adjacent portions of the spectrum is of prime importance for revealing details of the physical nature of the greenhouse effect, estimation of values of this effect, its variability, and effect on climate. Principally, a similar situation (but significantly different from the viewpoint of details) prevails on other planets (most serious efforts had been made till now only toward the study of the greenhouse effect on Venus and Mars [22]).

A major portion of the energy of thermal radiation is confined to the 3.5–50 μm region of the spectrum where more than 95% of the energy from the outgoing thermal radiation is contained. The intensity spectrum of outgoing thermal radiation [22] and backscattering of the earth's

---

*This section was written with the assistance of N.I. Moskalenko.

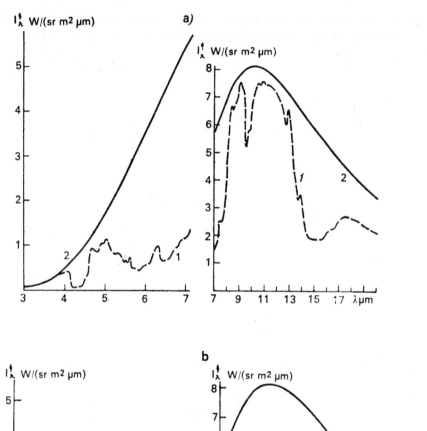

Fig. II.1. 1—spectral intensity of a) outgoing thermal radiation, and
b) backscattering of atmosphere; and 2—spectral brightness of an absolutely
black body at the surface temperature of the planet.

atmosphere, for two models based on its structural parameters, are presented in Fig. II.1. For many sections of the spectrum atmospheric back-
109 scattering compensates for the surface emittance. Only in spectral windows, namely 3.8; 4.8; 8–14 and 18 μm, the emittance of the underlying surface exceeds the backscattering of the atmosphere and thermal radiation into space is not fully compensated by the backscattering of the atmosphere. This has a significant influence because of variations in optical properties of the atmosphere in spectral windows on the radiant heat exchange and climate.

From the viewpoint of the theory of climatic variations the study of the effect of climate-forming factors with long-term trends is most important. In the investigations of the mechanisms of the greenhouse effect it is essential to consider changes of the optical properties of the atmosphere, which lead to variations in fluxes of outgoing radiation and backscattering and a change in the shielding action of the atmosphere. In this connection in [22] detailed computations of the effect of the shielding action of the earth's atmosphere with the consideration and analysis of all factors were carried out. These factors have a practical influence on the fluxes of upward and downward radiations, and it has been shown that the effect of minor components and atmospheric aerosol on the greenhouse effect was earlier underestimated.

It should be mentioned that the greenhouse effect of the atmosphere appears for the earth-atmosphere system only when the temperature of the surface of the planet exceeds that of the surrounding atmosphere. Otherwise, in the opposite case, the atmosphere will generate additional radiation in the bands of atmospheric absorption and create an antigreenhouse effect, intensifying, in this way, the radiational cooling of the planet. Such situations are observed for cold belts in the earth's atmosphere (in the Arctic and Subarctic), and also under conditions of a cloudy atmosphere [22]. For illustration, the spectrum of outgoing radiation computed from the model of the cold atmosphere of the earth is presented in Fig. II.2, in which the emission bands of the atmosphere are observed on a general background of underlying surface radiation.

In Fig. II.3 the values of temperatures computed from the spectral distribution of outgoing thermal radiation with the consideration of radiation absorption by water vapor (curve 1) and the actual chemical composition of the atmosphere (curve 2) are presented. It is seen that the minor components have the strongest impact in the region of the spectral window of 7–14 μm. This is why the study of optical properties of the atmosphere in the spectral window of 7–14 μm is of significant importance in explaining the optical characteristics of the atmosphere in the greenhouse effect and consequently in the global climate.

Many processes taking place in the atmosphere have a close correla-

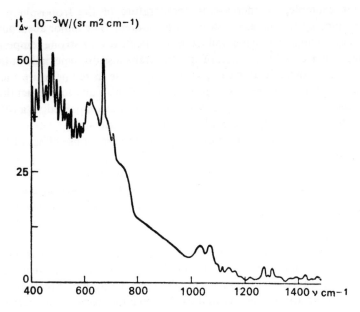

Fig. II.2. Spectral intensity of outgoing thermal radiation calculated for the model of atmosphere above the Arctic belt of the earth.

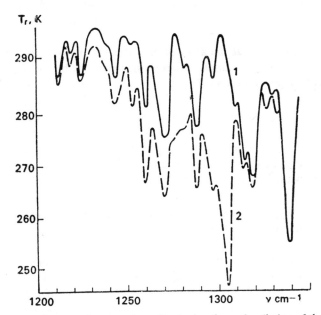

Fig. II.3. Spectra of temperature of outgoing thermal radiation of the earth, calculated 1) with consideration of absorption only by water vapor and 2) for actual chemical composition of the earth's atmosphere.

104

tion; for example, an increase in temperature of the troposphere is
110 accompanied by an increase in its moisture content. The major absorbing
components (carbon dioxide and ozone) indicate a very strong temperature dependence of the function of spectral transmission, especially at the
edges of the bands. In this case, with an increase in temperature their
absorption capacity increases considerably. All this leads to the fact that
an increase in temperature of the earth's surface is accompanied, as a rule,
by an intensification of the greenhouse-effect mechanism.

The traditional approach to the study of the greenhouse effect of the
earth's atmosphere and, in particular, its variations, are related with the
manifestation of the contribution of carbon dioxide [7]. One of the most
popular hypotheses in regard to climatic variations in the geologic as well
as the recent past, links the variation of temperature with changes in the
carbon-dioxide content of the atmosphere, causing changes in the greenhouse effect. However, the increasing scales of atmospheric pollution
cause increases in the concentration of such gaseous components as sulfur
dioxide, carbon monoxide, carbon halides, nitrogen oxides, nitric acid
vapors, hydrocarbon compounds, etc. The consideration of direct aerosol
pollution in the atmosphere as well as that resulting because of chemical
transformations of gaseous components into the solid phase (this relates
111 primarily to sulfate-bearing aerosol) is of significant importance. All
these components (including aerosol) possess absorption bands in the
infrared region of the spectrum, which determine their contribution to
the greenhouse effect of the atmosphere.

It is well known that the absorption spectrum of the atmosphere in
the region of a spectral window is determined not only by the effect of the
15 μm wavelength of carbon dioxide, but also by such gaseous components as water vapor, ozone and many other minor components, as well as
aerosol. Naturally in this case from the standpoint of the theory of
climatic variation, the most important role is assigned to those optically
active components which show long-term trends. In this connection
chlorofluorocarbons occupy a significant place, comprising that component of the atmosphere which possesses an exclusively anthropogenic
origin, particularly in view of increasing magnitudes of these being
released into the atmosphere (explained at a subsequent stage).

Starting from 1973, when Lovelock [116] had for the first time drawn
attention to the serious requirement of studying chlorofluorocarbons in
112 connection with their possible destructive impact on the ozone layer, and
till very recently, in spite of the great interest shown toward the problems
of chlorofluorocarbons by researchers, only the work of Ramanathan
[137] has been published where the study of the transfer of long-wave
radiation with consideration of chlorofluorocarbons has been taken up.
Presently this gap is being filled. Hence not only the necessity but also

the possibility has arisen of discussing the contribution of chlorofluoro-carbons in the greenhouse effect of the atmosphere.

Apart from the aforementioned trace elements, the atmosphere contains a complete array of others, such as nitrous oxide, methane, ammonia, nitric acid, ethylene, and sulfur dioxide, which possess absorption wavelengths in the 7–14 μm band. These contribute to the greenhouse effect of the atmosphere, especially if we consider that the concentrations of these components undergo severe changes due to the influence of anthropogenic actions (intensive application of organic fertilizers. burning of fuels, etc.).

The main conclusion drawn by Ramanathan [137] related to the need to take into account the contribution of chlorofluorocarbons to the green-house effect. This conclusion entirely corroborates the conclusions of Wardle and Evans [176]. Here, it has independently been shown that Freon pollution of the atmosphere, bringing in changes in the radiation budget of the earth-atmosphere system, can affect the climate more than any other anthropogenic factor (such as increase in concentrations of carbon dioxide and aerosol). In this work, while using a very simple model of climate, an important concept has been proposed: although the danger of the chlorofluorocarbon-ozone cycle is still not sufficiently understood, the effect of chlorofluorocarbons on the radiation budget must be assessed without delay while taking into consideration their observed low concentration.

The calculations carried out by Wang et al. [86] (the main results are represented in Table II.4) showed that a doubling of the concentration of nitrous oxide increases the average temperature of the earth's surface by 0.7 K. (The long-term changes of temperature by more than 0.1 K can be considered potentially significant, whereas a change of approximately 1 K reflects a fundamental variation in climate.) Doubling of methane and ammonia concentrations should lead to a cumulative increase of temperature by 0.3 and 0.1 K respectively. In Table II.4 the results of calculations of possible changes in the temperature of the earth's surface, caused by variations in the concentrations of HNO_3, C_2H_4, SO_2, CCl_2F_2, CCl_3F, CH_3Cl and CCl_4 concentrations are also presented. For the sake of comparison, the temperature variations due to water vapor, carbon dioxide and ozone are also given. It is important that the total green-house effect of doubling of N_2O, CH_4, NH_3 and HNO_3 concentrations attain 1.2 K. The effect of chlorofluorocarbons also becomes noticeable if their concentration increases by an order of magnitude.

114   Quantitative discrepancies in the results of computations obtained in [137] and [176] should now be considered; the greenhouse effect due to chlorofluorocarbons according to [176] is considerably lower than it was found in [137] (0.9 K). Nevertheless, the total quantitative assessment,

Table II.4. Additional greenhouse effect arising from an increase in concentration of various minor components of the atmosphere

Component	Center of the band, μm	Assumed present-day concentration[1], ppm	Increase of concentration by a factor	Greenhouse effect, K	
				at fixed temperature of upper boundary of clouds	at fixed altitude of upper boundary of clouds
N_2O	7.78; 17.0; 4.5	0.28	2	0.68	0.44
CH_4	7.66	1.6	2	0.28	0.20
NH_3	10.53	6×10^{-3}	2	0.12	0.09
HNO_3	5.9; 7.5; 11.3; 21.8	4.85×10^{-3} mm STP	2	0.08	0.06
C_2H_4	10.5	2×10^{-4}	2	0.01	0.01
SO_2	8.69; 7.35	2×10^{-3}	2	0.03	0.02
CCl_2F_2	9.13; 8.68; 10.93	1×10^{-4}	20	0.54	0.36
CCl_3F	9.22; 11.82	1×10^{-4}	20		
CH_3Cl	13.66; 9.85; 7.14	5×10^{-4}	2	0.02	0.01
CCl_4	12.99	1×10^{-4}	2		
H_2O	6.25; $10 - \infty$	Relative humidity 75%	2^2	1.03	0.65
CO_2	15.0	330	1.25	0.79	0.53
O_3	9.6	3.43 mm STP	0.75	−0.47	−0.34

[1]These ratios of mixture correspond to the level of the earth's surface, and for O_3 and HNO_3 the total content in the atmospheric layer has been indicated.

[2]It is assumed that there is an increase in H_2O content by two times above 11 km, and below this altitude, the moisture content is determined by the conditions of fixed relative humidity. The greenhouse effect caused by doubling of CH_2Cl_2, $CHCl_3$, C_2H_6 and PAN concentrations is negligibly weak (less than 0.01 K).

provided by Ramanathan [137], remains true; a sizable increase of Freon concentration in the atmosphere must have an impact on the climate.

With the aim of studying the impact of gaseous and aerosol components of the atmosphere on the greenhouse effect the calculations of thermal radiation fluxes in the atmosphere for models of the atmosphere with different chemical compositions were carried out in [22]. The computation scheme includes all the components having an effect on radiation absorption in the earth's atmosphere including water vapor (with the consideration of the continuum absorption), carbon dioxide, nitrogen and oxygen (pressure-induced absorption), methane, nitrogen oxide, nitrous oxide, nitrogen dioxide, sulfur dioxide, nitric acid vapors, ethylene, acetylene, ethane, formaldehyde, chlorofluoromethanes, ammonia, and aerosol formations of different chemical composition and microstructure. Data on the functions of spectral transmission obtained from direct calculations and on the basis of laboratory measurements has been used in the computation. For the basic components the effect of temperature on the function of spectral transmission has been taken into account. The conditions of clear, misty and cloudy atmospheres are considered. In accordance with the method described in Section 2, the entire input information was fed into a computer in a numerical form which enabled the modeling of the atmosphere not only of the earth, but also of other planets (Mars, Venus, Jupiter, Saturn).

The optical properties of the earth's atmosphere are well known. The major absorbent components are water vapor and carbon dioxide, which determine the radiation regime of the troposphere. Minor components (such as ozone, methane, nitrous oxide, sulfur dioxide, chlorofluorocarbon and nitrogen dioxide) have a significant effect on the transfer of radiation only in the spectral windows of water vapor and carbon dioxide. The spectral structure of these gases is very well known at present. The role of these minor components in the formation of the radiation regime increases in the stratosphere. In the computation of the greenhouse effect it is most important to take into account the radiation absorption due to the vibrational- and rotational line at 6.3 μm and the rotational line at $\lambda > 15$ μm for water vapor, due to the 4.3 and 15 μm lines of carbon dioxide, the 4.75 and 9.6 μm lines of ozone, the 4.5, 7.8 and 8.6 μm lines of N_2O; the 3.3 and 7.6 μm lines of CH_4, pressure-induced absorption of emission in the 4.3 μm line of N_2 on the N_2–N_2 and N_2–O_2 collisions of molecules and the 6.5 μm line of oxygen for the collision of O_2–O_2 molecules. The role of atmospheric vibrational- and absorption lines at 4.67 μm for CO and 5.3 μm for NO; the 6.3 and 10.6 μm lines for NH_3; and the 4, 7.5, 8.6, and 20 μm lines for SO_2 in the troposphere is not very substantial. In the stratosphere it is important to consider the transfer of radiation in the absorption lines, viz., the 5.6, 11.2 and 20 μm lines of HNO_3; the 5.5, 7.35 and 8.69 μm lines of NO_2; the 9.22 and 11.82 μm lines of $CCIF_3$; the 8.68, 9.13 and 10.93 μm lines of CCl_2F_2; the 7.14, 9.85 and 13.66 μm lines of $CHCl_3$ and the 12.99 μm line of CCl_4.

Many of the aforementioned gaseous components which absorb radiation in the spectral region of thermal radiation are products of industrial activity and, therefore, their concentration will only increase with time. In this regard the example with chlorofluorocarbons [20, 21] is indicative of this effect. If, at present, the volumetric concentration of F_{11}, F_{12} is about $(1-2) \times 10^{-10}$, then by the end of the century at the present rate of atmospheric pollution, the concentration of chlorofluorocarbons may increase to 2×10^{-9}. The longer lifetime of F_{11} and F_{12} in the atmosphere promotes the process of stratospheric Freon accumulation. It is anticipated that by the end of the century the concentration of carbon dioxide may become twice as much as it is at present.

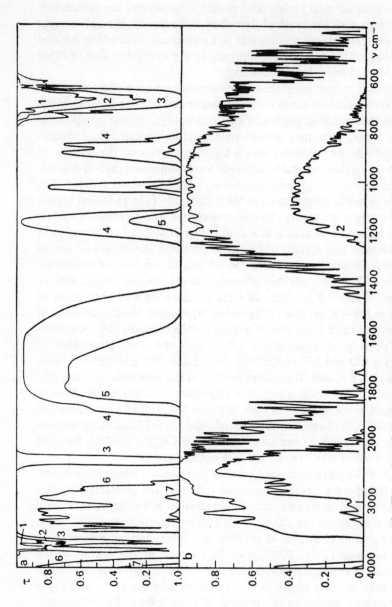

Fig. II.4 (a, b). Absorption spectra of some gaseous components of planetary atmosphere:
a—carbon-dioxide content, atm cm: 1) 0.63, 2) 2.5, 3) 10, 4) 10⁴, 5) 21×10⁴, 6) 4×10⁴, 7) 14×10⁴, pressure, mb;
1) 63, 2) 250, 3) 10³, 4) 10⁴, 5) 10⁴, 6) 10⁴, 7) 10⁴; b—water-vapor content, g/cm²: 1) 0.2, 2) 10; pressure, 10³ mb, pH₂O=20.

109

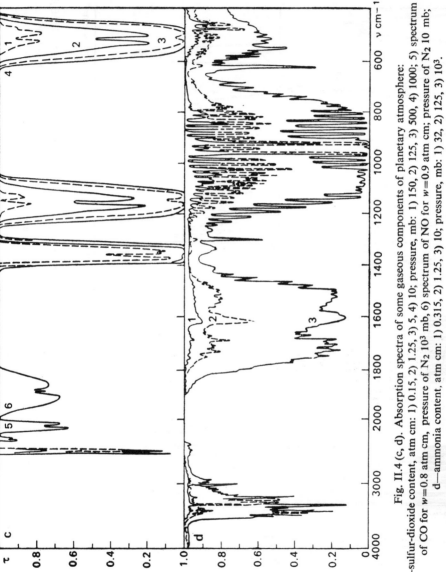

Fig. II.4 (c, d). Absorption spectra of some gaseous components of planetary atmosphere:
c—sulfur-dioxide content, atm cm: 1) 0.15, 2) 1.25, 3) 5, 4) 10; pressure, mb: 1) 150, 2) 125, 3) 500, 4) 1000; 5) spectrum of CO for w=0.8 atm cm, pressure of N_2 10^3 mb, 6) spectrum of NO for w=0.9 atm cm; pressure of N_2 10 mb;
d—ammonia content, atm cm: 1) 0.315, 2) 1.25, 3) 10; pressure, mb: 1) 32, 2) 125, 3) 10^3.

117

110

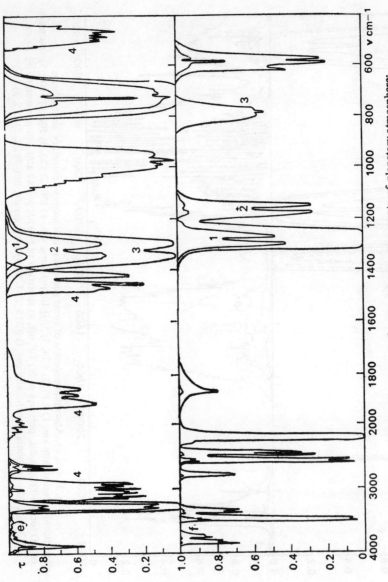

Fig. II.4 (e, f). Absorption spectra of some gaseous components of planetary atmosphere: e—acetylene content, atm cm: 1) 0.16, 2) 2.5, 3) 10; pressure, mb; 1) 16, 2) 250, 3) 10³, 4) curve for ethylene with C₂H₂ content of 2.2 atm cm at pressure of 10³ mb; f—nitrogen-oxide content, atm cm: 1) 0.31, 2) 10, pressure, mb; 1) 31, 2) 10³, 3) absorption spectrum of ethane for acetylene content of 1.2 atm cm.

118

111

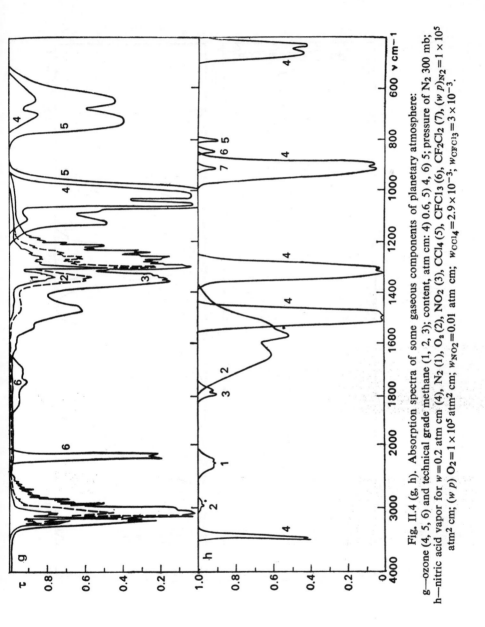

Fig. II.4 (g, h). Absorption spectra of some gaseous components of planetary atmosphere:

g—ozone (4, 5, 6) and technical grade methane (1, 2, 3); content, atm cm: 4) 0.6, 5) 4, 6) 5; pressure of N_2 300 mb; h—nitric acid vapor for $w = 0.2$ atm cm (4), N_2 (1), O_3 (2), NO_2 (3), CCl_4 (5), $CFCl_3$ (6), CF_2Cl_2 (7); $(w\,p)_{N2} = 1 \times 10^5$ atm^2 cm; $(w\,p)$ $O_2 = 1 \times 10^5$ atm^2 cm; $w_{NO2} = 0.01$ atm cm; $w_{CCl4} = 2.9 \times 10^{-3}$; $w_{CFCl3} = 3 \times 10^{-3}$.

119

112

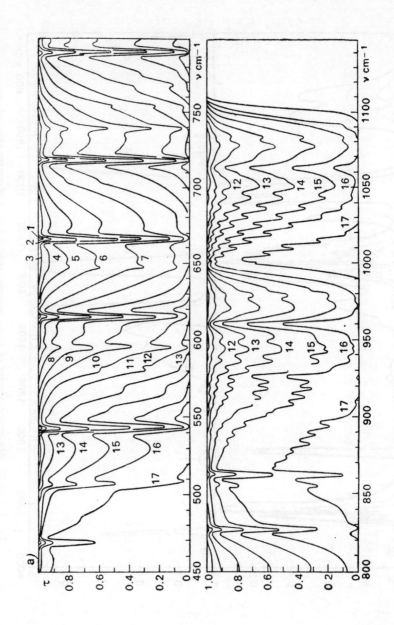

113

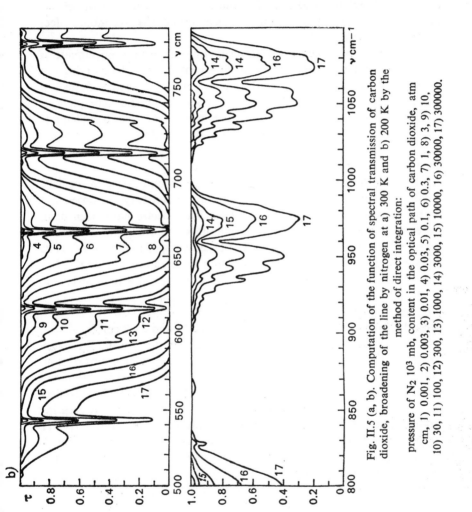

Fig. II.5 (a, b). Computation of the function of spectral transmission of carbon dioxide, broadening of the line by nitrogen at a) 300 K and b) 200 K by the method of direct integration:

pressure of N_2 10^3 mb, content in the optical path of carbon dioxide, atm cm, 1) 0.001, 2) 0.003, 3) 0.01, 4) 0.03, 5) 0.1, 6) 0.3, 7) 1, 8) 3, 9) 10, 10) 30, 11) 100, 12) 300, 13) 1000, 14) 3000, 15) 10000, 16) 30000, 17) 300000.

121

The absorption spectra of various gaseous components of the atmosphere from the experimental investigations by Moskalenko et al. [22] are presented in Fig. II.4. These spectra reflect the structure of transmission functions for selected conditions of observation. It is important to note that the absorption lines of various components of the atmosphere overlap. This shows the necessity of simultaneously considering the effects of many gaseous components on the mechanism of the greenhouse effect in the atmosphere.

Assessment of the effect of atmospheric composition on climate is carried out within the framework of an accepted model by carrying out numerical experiments. This enables us to reveal the degree of effect of various factors and trace and tendency of possible climatic consequences. In spite of the limited application of such assessments due to inadequate considerations of cause-and-effect relationships and the dynamics of climate-forming processes, the numerical experiments enable us to ascertain the tendencies of climatic changes in the future.

The transmission spectra of atmospheric carbon dioxide in the 450–1150 cm^{-1} region for a pressure of 1013.3 mb and varying contents of carbon dioxide are shown in Fig. II.4b. The spectra have been computed from the parameters of the carbon-dioxide lines by the method of direct integration for the conditions of broadening of the carbon-dioxide lines by nitrogen (resolution 2.5 cm^{-1} at a temperature $T = 300$ K). For spectral lines in this case the Lorentz contour is applied with a correction:

$$b'(v - v_i) = \exp\left[-a(|v - v_i|\gamma - d)^c\right],$$

with the following parameters: $a = 0.44$, $c = 0.61$ and $d = 3$ cm^{-1}.

The results of similar calculations for temperature $T = 200$ K are presented in Fig. II.5. The fact, that with an increase in carbon-dioxide content the spectral optical density increases and boundaries of vibrational-and-rotational bands expand, demonstrates that the greenhouse effect caused by carbon dioxide is continuously intensified with a decreasing transparency of the atmosphere. The relationship of temperature rise ΔT on the earth's surface with an increase in carbon-dioxide concentration (as compared with its contemporary concentration), illustrating this conclusion, is presented in Fig. II.6.

The temperature of the earth's surface increases monotonically with an increase in carbon-dioxide concentration in the atmosphere. Doubling of this concentration would result in a temperature rise, for example, by 2.2 K. Furthermore, an increase in carbon-dioxide concentration by 4 and 10 times would lead to a rise in surface temperature by 4 and 6.5 K respectively.

In the calculations of increase in temperature of the earth's surface with an increasing carbon-dioxide concentration, the vertical profiles of atmospheric temperature were not taken into account. Consideration of the latter leads to higher values of ΔT. It is also important to bear in mind the fact that with a rise in the mean temperature of the earth's surface, the concentration of water vapor also increases, and consequently the opacity and warming the earth's atmosphere is intensified. Further, any increase in the mean temperature of the troposphere leads to an increased opacity of the atmosphere as a consequence of the severe effect of temperature on the transmission function of carbon dioxide.

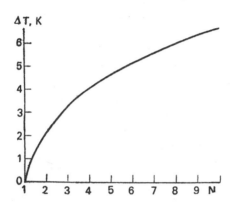

Fig. II.6. Greenhouse effect due to increased carbon-dioxide concentration in the atmosphere by N times as compared to that observed at present.

Insofar as the lower stratosphere is colder than the troposphere, the mechanism of the greenhouse effect as a result of a change in chemical composition is revealed more effectively in this case. Thus, an increase in the quantity of water vapor in the stratosphere to twice as much leads to a warming by 1 K and a decrease in ozone concentration by 25% due to the destructive impact on the ozone layer leads to an antigreenhouse effect constituting (0.4–0.5) K.

123   The results mentioned here have been obtained for a mean global model of the atmosphere with fixed temperatures of the upper boundaries of clouds or fixed altitude of their upper boundaries. Apart from this, in a real atmosphere, changes in the concentration of gases are also accompanied by variations in their vertical profiles depending on the area of observation. For these reasons, the results obtained above on the effects of concentration variations of gaseous components on the greenhouse effect should be viewed as approximate.

The effect of aerosol on the radiation regime of the atmosphere occurs due to scattering and absorption of radiation by aerosol [18]. In this, depending on the albedo of single scattering, conditions are created in which the effect of aerosol either increases or decreases the earth's albedo.

116

[18]. Here the ambiguity in the conclusions is determined by an uncertainty of the optical properties of aerosol caused by a wide space-time variability in the concentration and microstructure of aerosol, and also because of the relationship of the chemical composition of aerosol with the conditions of its generation.

The fact that the capacity of aerosol to heavily absorb and scatter radiation depends on its microstructure, is important. The aerosol attenuation factors for the continental dust fraction of aerosol as calculated by Moskalenko [22] are presented in Fig. II.7. The trends of attenuation, absorption and scattering factors for fine aerosol particles (Aitken nuclei) should be given special attention. Even for $\lambda = 0.55$ μm, the absorption factor σ_{as} constitutes only 30% of the overall attenuation factor σ_a. For $\lambda > 1$ μm, the contribution of σ_{as} to the overall attenuation factor σ_a is less than 1%. In the 6–12 μm spectral range, the attenuation of a submicronic fraction of aerosol σ_a increased because of an increase in the imaginary part of the refractive index and was determined by the absorption of radiation by aerosol. For larger particles, scattering provides a predominant contribution to the total attenuation in the $\lambda < 7.5$ μm spec-

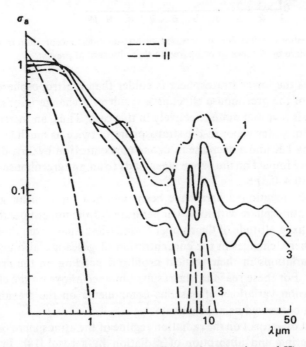

Fig. II.7. Spectral dependence of I) coefficient of attenuation and II) scattering for model of 1) fine-dispersed fractions of aerosol and of 2, 3) aerosol mist of medium dispersion.

tral wavelength. It is a characteristic feature that, at the same optical density (visibility range) of aerosol, the absorption factor for finer particles is higher than for larger particles, $\lambda = 0.55\ \mu m$. Such spectral distribution of the attenuation factor is explained by the spectral dependence of the imaginary part of the refractive index $\varkappa(\lambda)$. For small particles the attenuation factor σ_a is proportional to $\varkappa/\lambda$). Increase of $\varkappa$ in accordance with an increase in λ arises because of an almost direct proportionality, viz. $\sigma_a(\lambda) \approx \sigma_{as}$.

Hence, fine-dispersed aerosol, while intensely absorbing solar radiation, warms the atmosphere and emits in the long-wave region of the spectrum shielding the thermal emission of the earth's surface. Aerosol, as an absorbent gaseous component, intensifies the greenhouse effect of the
124 atmosphere through long-wave radiation and increases, for mean global atmospheric conditions, the temperature of the earth's surface by approximately 3 K. An increase in the concentration of tropospheric aerosol by 1.5 times due to industrial activity would lead to a warming by 1.7 K. No less important is the consideration of the stratospheric aerosol layer in the mechanism of the greenhouse effect. Doubling of the opacity of stratospheric aerosol leads to a rise in temperature by 0.8 K.

In assessing the effect of aerosol on radiant heat exchange, the possible temperature differences of aerosol particles, caused by absorption of solar radiation, from the equilibrium temperature of profiles of the atmosphere were not considered. The higher temperature of aerosol particles as compared with the temperature of gaseous components can cause an increase in radiational cooling of the atmosphere and reduce the effect of stratospheric aerosol on the greenhouse effect.
125 The present-day industrial activity of man has already led to considerable reduction of the vegetation cover on the earth and an increase in the effect of mineral and industrial aerosols on the radiation regime of the atmosphere [21, 27, 28]. However, for the purpose of forecasting the aerosol effect on climate it is important to have an overall account of the anthropogenic impact, geochemical processes, physico-chemical processes occurring in the atmosphere and the biosphere, and turbulent heat exchange in the atmosphere. In this, an important role is played not only by an increase in density of the aerosol layer but by the change in its chemical composition, microstructure and spatial structure.

From the results cited here, there is no doubt about the importance of the climatic effect of a number of trace elements, the effect of which on climate has been assumed earlier as negligibly weak. On the other hand, from this follows the conclusion regarding the necessity of simultaneous monitoring of global trends of concentration of such trace elements. Undoubtedly, however, factors requiring immediate attention would include the increase in concentration of carbon dioxide, chlorofluoro-

methanes and also the variations of water-vapor content in the stratosphere. The results obtained indicate a probable warming of climate,
caused by anthropogenic production of minor gaseous components, which
should have special importance for the continental regions in temperate
and higher latitudes of the northern hemisphere. Apparently, aerosol cooling cannot compensate for the effect of warming of climate. However, reliable quantitative assessments are possible only on the basis of the use of
more perfect models of the climate and a comprehensive account of the
contribution of the aerosol effect. In this connection the investigations on
climate-forming factors on other planets are of great importance.

Results obtained by researchers demonstrate the fact that the generally accepted hypothesis regarding the origin of contemporary climatic
changes and possible factors and tendencies of future climate, which
attribute fluctuations of climate only to variations in the carbon-dioxide
content (especially in the future), are nothing but a rough schematic
representation of reality. Even the 'carbon-dioxide' hypothesis is laden
with a number of limitations.

Insofar as the anthropogenically caused changes of atmospheric composition depend on several factors, it is natural that theoretical assessments of their impact on the greenhouse effect and climate should take
into account the most significant components. In this regard, Table II.4
126 may serve as a guide. The specific conclusions, related to this, on the role
of various components should necessarily be considered in designing a
system for observing the important (from the viewpoint of predicting
climate) parameters and their determining factors. Nevertheless, the investigation of the absorption spectra of 'climatically significant' pollutants is of primary importance.

The various theoretical models of climate constitute a basis for an
assessment of the effect of atmospheric composition on climate. In this
connection attention should be drawn to the fact that numerical experiments on the sensitivity of climate to different parameters (for example,
carbon-dioxide concentration) should not be considered as a sequential
cause-effect relationship. In order to improve the reliability of such
assessments, consideration of the actual dynamics of continuous changes
in the atmospheric composition in all its diversity (instead of, for example,
an arbitrary assumption about doubling of the carbon-dioxide concentration) is of great importance. Even 'complete' three-dimensional models
require modifications for the explanation of 'mechanisms' responsible for
climatic variations, especially from the viewpoint of an adequate account
of interactions of the atmosphere, ocean and the cryosphere.

3. COMPOSITION OF THE STRATOSPHERE AND ITS DETERMINING FACTORS

Ozone appears to be optically the most active gaseous component in the stratosphere. The problem of shortwave radiation absorption by ozone and its effect on the transfer of radiation has been extensively discussed in the literature [2, 4, 5, 8, 10, 108]. As mentioned earlier, during recent years the problem of anthropogenic impact on the ozone layer has added new dimensions, which provided an impetus to review the theory of the ozone layer, and led to new results and impelling scientists to assess afresh the role of ozone as a climate-forming factor. In view of further developments in the theory of the ozone layer, it has become necessary to widen investigations to cover all gaseous components responsible for ozone formation in the stratosphere.

3.1. Introductory remarks. Investigations carried out during recent years showed that the 'ozone shield', which protects all living beings on the earth from the deadly effects of severe ultraviolet radiations and has significant importance from the viewpoint of climate formation, is highly sensitive to external influences. Although the initial threatening forecasts of catastrophic destruction of the ozone layer by supersonic jet aircraft appear to be incorrect, the establishment of an extensive Climatic Impact Assessment Program (CIAP), which was responsible for the publication of six monographs [57], conclusively showed that there must be a definite impact of supersonic jet aircraft on the ozone layer and these effects are caused primarily by oxides of nitrogen. Subsequent developments have directed attention to the problem of the impact of chlorofluoromethanes injected into the atmosphere in large quantities on ozone, and related with this, the 'chlorine' mechanism of breakdown of the ozone layer [30, 34, 83]. Associated with these new developments, the modifications in certain reaction coefficients have radically changed initial conclusions.

Since the problem of oxides of nitrogen was widely discussed in the literature and appeared to be considerably less important than was earlier anticipated [45, 46, 68, 94–97], we shall limit ourselves mainly to discussion of the chlorine mechanism. It should be noted that although we will not consider biological and other consequences of a decrease in atmospheric ozone content, these are of exceptional importance [57].

Discussion on the effect of nitrogen oxides. In referring to the problems of oxides of nitrogen it need only be mentioned that till now discussion regarding the validity of results from observational data, endorsing their effect on the ozone layer is continuing. In particular this relates to some such confirmations of results from observational data. Ellsaesser [73, 79] considers, for example, that a sudden warming of the stratosphere (SWS)

is indicative of the inadequacy of the hypothesis of catalytic destruction
of ozone by oxides of nitrogen.

Usually it is assumed that a sudden warming of the stratosphere
occurs as a result of an adiabatic warming (reaching up to 80 °C over
several weeks) which is caused by the descent of a considerable portion
of the stratosphere by 10–20 km in the latitudinal zone above 40°. Warming of the stratosphere at higher latitudes is accompanied by its cooling
in the tropics of the other hemisphere. A similar compensation also takes
place along the vertical axis with a characteristic wavelength of ~45 km.
All this shows that SWS is a phenomenon on a global scale. The necessity
for vertical movements accompanying SWS determines the increased
concentration of oxides of nitrogen in the heated regions of the stratosphere.

In spite of the aforementioned, several months after a sudden warming of the stratosphere, a maximum ozone content is observed. It is maximum in the northern hemisphere where the most severe warming takes
place. In [73] statistical data on SWS and ozone is presented, which
128 indicates a distinct increase in total ozone content north of 40°N accompanying SWS. An intense SWS appears to be the major factor determining
the variability of total ozone content from one year to the next.

Ellsaesser [77] expressed his views on the inaccuracy in the interpretation of observational data from the *Nimbus-4* satellite, according
to which a decrease in ozone content by 16% was observed after solar
flares in August 1972. These observations cannot, therefore, be considered as a confirmation that photochemical phenomena (including
those related to release of oxides of nitrogen by supersonic jet aircraft)
can cause a sharp and significant decrease in the ozone content. The
results of rocket measurements during November 2–4, 1969, in the
period of a proton event revealed a decrease in ozone concentration at
an altitude of 54 km to half and, at 67 km, to a quarter of its value at the
moment of maximum intensity of the proton event. These results were
stated as a proof of the theory of catalytic destruction of ozone by the
HO_x-type radicals. According to Ellsaesser, the contradiction in the interpretation of identical facts demonstrates the absence of an adequate basis
for a reliable judgement of the factors determining the variability of the
ozone content.

In a letter [145], containing statements against the views of Ellsaesser,
an argument was presented in favor of the fact that a decrease in the
ozone content, detected from data of rocket measurements at the time
of a proton event during November 2–4, 1969, in the mesosphere (at
altitudes of 54 and 67 km), actually proves the theory of catalytic destruction of ozone by HO_x-type radicals. Above 50 km it is this mechanism, and not the oxides of nitrogen, which is responsible for destruction

of ozone. Decomposition of ozone by the latter takes place only below 45 km and here the restoration of the ozone content appears to be much slower as compared to that in the mesosphere. This confirms the results of ozone measurements from the *Nimbus-4* satellite.

These results indicate a spontaneous decrease in the ozone content in the stratosphere (above 30 km) at higher latitudes during the period of the proton event in August, 1972. Furthermore, the decreased ozone content was noticed during the course of 2–3 weeks after the event. The calculated decrease in ozone content agreed with the values measured within an accuracy of a few percentages. For altitudes below 30 km neither computations nor observations indicated such a decrease, since a proton event did not produce at these altitudes, significant quantities of NO. At the same time, 2–3 weeks is too short a period for a transfer of NO downward from the upper layers of the atmosphere. All this demonstrates that there is no other way to explain the impact of a proton event on stratospheric ozone, except the application of the photochemical theory which takes into account the catalytic destruction of ozone by oxides of nitrogen. Undoubtedly, the modern photochemical theory which takes the diffusion processes into account requires further development.

Numerical models of the stratosphere. Summing up the accomplishment of CIAP, Hidalgo [94] stressed, that a correct solution of the problem regarding the effect of supersonic jet aircraft on the ozone layer is possible, only if the complete system of equations defining the thermo- and hydrodynamics of the anthropogenic impact is used. The system of equations is supplemented by equations of the kinetics of minor components of the atmosphere, which characterize the variations in chemical composition of the atmosphere caused by processes such as phase transformation, precipitation, adsorption, chemical reactions, etc. In such formulations of problems, the specific features of interaction between thermo- and hydrodynamics and chemical transformations determining the changes in composition of the atmosphere, can be explained more comprehensively.

The list of chemical reactions discussed in monographs devoted to the results of CIAP [57] contains 130 reactions (see [142] also). These reactions should be taken into account in any numerical model describing the processes in the stratosphere. The number of these reactions can be reduced to 55 most important reactions or even 34 reactions, in which 17 components participate, including reactions of the following cycles: O_x [O_3, O_2, O (^1D), O (^{3}P)], NO_x [N_2, N, N_2O, NO, NO_2, NO_3, HNO_3, HNO_2] and HO_x [H_2O, H, HO, HOO, H_2O_2]. In recent years the decisive nature of reactions of the chlorine cycle ClO_x and correctness of the data on reaction rates has been explained.

Hidalgo [96] showed that the modification of a set of photochemical reactions and rates of a number of reactions led to the conclusion on the change even in the sign of anthropogenically caused variations of the ozone content. According to the data obtained from computations, carried out by Hidalgo and Crutzen [95] without consideration of the chlorine cycle, the impact of supersonic jet aircraft on ozone leads to a situation where in the zones of intense flights of subsonic aircraft at an altitude of 10.8 km a slight increase (less than 1%) in the total ozone content has been noticed, on account of reactions of methane oxidation. However, in the case of release of oxides of nitrogen at an altitude of 14.5 km there is practically no effect, and releases at a level of 18 km lead to some decrease in the ozone content (under the influence of oxides of nitrogen). Subsequent estimates led to the conclusion regarding an increase in the ozone content in the upper stratosphere.

In the most general form the problem of a natural layer of ozone and 130 the anthropogenic impact on it are solved on the basis of a complete system of equations of the thermo- and hydrodynamics and kinetics of gaseous components of the stratosphere:

$$dU/dt = -f\mathbf{k} \cdot \mathbf{U} - g\,\Delta z + \mathbf{F}; \tag{II.1}$$

$$dT/dt = kT\,\omega/p + Q/c_p; \tag{II.2}$$

$$\nabla \mathbf{U} + \partial\omega/\partial p = 0; \tag{II.3}$$

$$\partial z/\partial p = -RT/gp; \tag{II.4}$$

$$dR_i/dt = P_i - L_i. \tag{II.5}$$

Here U is the horizontal velocity vector of air, t the time, $f\mathbf{k}$ the vertical component of the vector of angular velocity of the earth's rotation, gz the geopotential, p the atmospheric pressure, $\mathbf{F}$ the viscosity, T the temperature, $\omega = dp/dt$ the vertical component of velocity, Q the heat influx (this term describes the interaction of radiative, thermal and dynamic processes), c_p the specific heat at constant pressure, R the Universal Gas Constant, R_i the mixing ratio of the ith gaseous component, and P_i and L_i the sources and sinks of this component.

The work of Ramanathan and Grose [140] is an example of the application of development of the climate theory of the stratosphere without consideration of the photochemical processes. Considering the model of seasonal climate of the stratosphere, they proceed from the assumption that the most important external source of energy for the stratosphere is the solar radiation absorbed by the stratosphere (mainly by ozone) and the mechanical energy transmitted from the troposphere. The thermal and dynamic states of the stratosphere are determined primarily by its radiative and dynamic reactions on the energy sources.

The impact of radiation factors arises because of transfer of long-wave radiation in the water vapor, carbon dioxide and ozone bands and can be expressed in two ways: 1) as radiant heat exchange between various layers of the troposphere and lower stratosphere, and 2) as long-wave cooling in space which serves as an important mechanism for dissipation of vortexes and available potential energy appearing in the stratosphere under the impact of external sources of energy.

The amount of long-wave cooling in space is proportional to the Newtonian rate of cooling h (h^{-1} being the thermal relaxation time). It was earlier shown that h closely depends on the latitude, the period of the year, and altitude (latitudinal variations and annual trends are determined by the temperature relationship of h). As a result of such dependences there are corresponding fluctuations of radiative interaction between the stratosphere and the troposphere.

131   In [140] two types of numerical experiments, with the aim of assessing the impact of latitudinal variation and annual trend of radiative interaction on the zonal circulation in the stratosphere (zonal fields of temperature and wind), have been undertaken. In the first numerical experiment (FNE) the exact computation of variations of temperature in the stratosphere, caused by long-wave radiation (and taking into account the contribution of carbon dioxide, water vapor and ozone) has been achieved. In the second numerical experiment (SNE), the approximation of long-wave cooling in space by assuming the weighted global annual vertical profile h has been used. This signifies, that the radiative interaction between the lower stratosphere, and the troposphere, as also the latitudinal-and-seasonal variations of thermal relaxation time have been neglected.

In both cases, latitudinal and intrayear variability of the following factors have been equitably considered: 1) warming of the stratosphere because of solar radiation, 2) nonadiabatic influxes of heat into the troposphere, and 3) nonzonal perturbations in the troposphere. The numerical experiments under consideration enable a study of the effect of latitudinal fluctuations and the annual trend of long-wave radiant heat exchange on the corresponding variation of climate in the stratosphere and also a study of the applicability of the long-wave cooling approximation in space.

Numerical modeling is based on the quasi-geostrophic nine-level global model of general circulation of the atmosphere. The initial conditions assigned for FNE and SNE have been found by integrating equations over a period of 9 months, starting from the quiescent state. Following this, the integral was evaluated for a period of 3 months (for a fixed zenith angle of the sun) and the last 30 days of this period were considered as characteristic of a stable regime at the moment of equilibrium. In order to simulate the annual trend, integration over an additional period of 33 months was carried out. Of these, the last 12 months are considered to be descriptive of the annual trend. The last days of this cycle correspond to the period of solstice (winter solstice in the northern hemisphere). The stable regime of the solstice was determined by integrating over 3 months, considering the data for the last days as the initial data, and on averaging for the last 30 days of this period. The same procedure was followed for the FNE and SNE.

The testing of the model indicated that it can simulate stratospheric climate fairly close to reality up to the 1-mb level. For an assessment of the contribution of radiant heat exchange between the lower stratosphere and the troposphere and also of latitudinal seasonal variations, in [140], 132 the differences in the results of the FNE and SNE as applied to the data for the annual trend of mean zonal fields of temperature and wind have been analyzed.

The results, thus obtained, indicate that the latitudinal distribution of temperature in the lower stratosphere is determined by the combined effect of dynamic warming due to absorption of solar radiation by ozone and interaction between the troposphere and the lower stratosphere arising because of radiant heat exchange. Consideration of variations in this interaction with latitude leads to maximal variations (up to 5–10 K) of the meridional gradient of temperature at the level of the tropopause and the lower stratosphere during winter and spring.

At the level of the tropopause, the warming caused by ozone (by absorption of solar radiation, as well as radiant heat exchange) attains a maximum in temperate latitudes during winter periods. This warming along with the processes of dynamic warming supports the existence of a warm belt within the lower stratosphere in temperate latitudes. A drop in temperature at the level of the tropopause, north of the warm belt (in the latitudinal range above 60°), during winter and summer is related with a change in the sign of the radiant heat influx because of carbon dioxide and water vapor from a positive value at the equator to a negative in the polar zones.

The large seasonal and latitudinal fluctuations in h exert a considerable influence on the latitudinal temperature gradient and its annual trend in the middle and upper regions of the stratosphere, and lead to cooling of the polar upper stratosphere during winter and an increase in the 'pole-equator' temperature gradient during winter and spring. An analysis of the results obtained by varying the ozone concentration at the level of the tropopause, enabled us to draw the conclusions that rapid changes in the sign of the vertical temperature gradient at the level of the equatorial tropopause may be due to an increased vertical gradient of radiant heating above the tropopause caused by ozone.

The results of numerical modeling of processes in the stratosphere, discussed here, show that the long-wave (Newtonian) cooling approximation in space is unsuitable for simulation of the climate of the lower stratosphere. However, this approximation is suitable in the case of models in which the vertical temperature profile is considered. In view of this, a fairly firm applicability of long-wave cooling in space is possible only at pressure of less than 5 mb, if the latitudinal-seasonal variability of h is taken into account.

A comparison between results of numerical modeling indicates the contribution of such factors in the development of stratospheric processes, such as the radiant heat exchange between various layers of the atmosphere and the temperature dependence of the thermal relaxation time, h^{-1}. In [53] a detailed analysis of the impact of these factors on the energetics of the stratosphere (layers of the lower stratosphere, between 120–20 mb and the upper stratosphere, located above the level of 20 mb) has been undertaken. Calculations revealed the presence of an annual trend, with maximums during winter and minimums during summer, of such components of the energy budget as the vortex components of kinetic (K_E) and all-available potential (A_E) energy, the formation of all-available potential energy $G(A_E)$ and vertical components of fluxes of vortex geopotential energy, $V(\Phi_E)$ caused by the propagation of planetary waves from the troposphere into the stratosphere.

For zonal components of the kinetic K_z and all-available potential A_z energy, and the rate $G(A_z)$ of generation A it can be shown that the semi-annual periodicity with the principal maximum in winter, the secondary maximum in summer and corresponding minimums during spring and autumn are characteristic. The annual trend of $V(\Phi_E)$ is determined by the fact, that a winter transfer from west to east creates favorable conditions for the upward propagation of planetary waves, whereas a circulation of the opposite sign during summer, hinders this. The semi-annual periodicity of K_z, A_z and $G(A_z)$ is related with the seasonal fluctuations of circulation in the stratosphere: transition from the western transfer to the eastern in summer and from a cold winter polar stratosphere to a warm summer stratosphere.

A comparison of spectra of A_E and K_E enabled us to find a decrease in the all-available potential as well as kinetic energy with an increasing wave number, but the fall in A_E appears to be more rapid than in K_E. The analysis of data from numerical modeling of the energy cycle shows that $V(\Phi_E)$ is the main source of energy for the lower stratosphere, whereas $V(\Phi_E)$ and $G(A_z)$ provide the major contribution to the energetics of the stratosphere.

For reasons mentioned here, the annual trend of vortex components of the energy budget K_E, A_E and $G(A_E)$ is determined by the vortex geopotential energy $V(\Phi_E)$, while the seasonal fluctuations of zonal components are controlled by a latitudinal variability of heat influx. This appears especially in the upper stratosphere during summer, when $V(\Phi_E)$ decreases and the zonal state is determined by $G(A_z)$.

A complete consideration of radiant heat influx leads to considerably higher mean annual values of vortex kinematic and all-available potential energy (Table II.5). A comparative analysis of the results of computation relating to FNE and SNE shows that the consideration of radiant heat influx because of long-wave radiation manifests itself as follows: 1) the temperature relationship of the Newtonian rate of cooling h causes an increase of $G(A_z)$ and strengthens the annual trend of $G(A_z)$ in the upper stratosphere; 2) this temperature relationship also promotes an improvement in conditions for the propagation of planetary waves through the upper stratosphere during winter and autumn; 3) the radiant heat exchange between the troposphere and the lower stratosphere changes the profile of zonal winds in winter, which favors the propagation of planetary waves in the lower stratosphere; and 4) under the influence of the second and third factors mentioned above, the value of the vortex components of energy for the FNE is considerably higher than in the case of the SNE.

Table II.5. Mean annual values of kinetic and potential energy in northern (NH) and southern (SH) hemispheres (10^2 J/m^2)

Type of experiment	A_z		K_z		A_E		K_E	
	NH	SH	NH	SH	NH	SH	NH	SH
Upper stratosphere								
FNE	259	277	428	703	51	44	222	198
SNE	305	300	453	491	30	38	138	160
Lower stratosphere								
FNE	677	690	477	818	328	246	714	592
SNE	795	678	454	496	288	214	575	459

A complete three-dimensional model of general circulation in the atmosphere encompassing the layer of the atmosphere from the earth's surface to the stratopause has been developed in the Lawrence Livermore Laboratory of the University of California at Los Angeles [124]. In this model for the first time the interactions of circulation, radiative transfer and ozone photochemistry have been taken into account. The array of the main prognostic variables of the model of general circulation of the atmosphere (the basis of which are the models of Arakava and

Mintz) includes the horizontal component of velocity, temperature and ground pressure. Apart from these, the additional prognostic variables are moisture (this is linked with the consideration of parameterized convection providing for cooling), temperature of the underlying surface, moisture reserves of the soil/rock, mass of snow, and total ozone content. The most significant diagnostic variables are the amount of clouds of various types, surface albedo (taking into consideration the relationship between the moisture reserve of soil and vegetation, because of which it is possible to parameterize biogeophysical feedback links) and many others.

The block for formation and transformation of ozone takes into account the Chapman and nitrogen cycles (spatial distribution of nitrous
135 oxide is assumed). Further, the 12-level model of the atmosphere has the upper boundary at the level of 1 mb (about 50 km), while for the uppermost layer (or the 'lip') a term accounting for damping has been introduced, which ensures absorption of the energy of waves being propagated upward. The mean annual temperature over the ocean is assumed at the level of the underlying surface. The horizontal resolution is 5° longitude × 4° latitude. The time interval, as a rule, is equal to 5 min (heat influx as well as vertical advection of water vapor and ozone are calculated at 30 min intervals).

In [124] the results of preliminary numerical modeling relating to winter in the northern hemisphere are discussed. Insofar as the computations only cover a period of 76 days, it should not be considered that the results obtained characterize conditions of statistical equilibrium. Nevertheless the calculated fields of meteorological factors agree fairly well with the observations. It is assumed, therefore, that the portion of results which is not established from observational data can be considered as a zero-order approximation describing the actual atmosphere.

Analysis of computed charts of the total ozone content demonstrate the presence of considerable zonal fluctuations, especially at temperate latitudes of the northern hemisphere. The photochemical processes of ozone formation are characterized by the presence of, a) its sources in the tropics and temperate latitudes of the southern hemisphere, b) sinks at higher latitudes of the southern and temperate latitudes of the northern hemisphere, and c) by the fact that there is no production of ozone during the period of polar night in the northern hemisphere.

Meridional transfer of ozone in the layers of the tropical atmosphere is caused almost entirely by a large scale transfer from summer to winter hemispheres with the help of the mean meridional circulation. However, at temperate and higher latitudes it represents the small difference between large macroturbulent transfer directed toward the poles, and a large transfer toward the equator caused by the mean meridional circulation.

In its turn the large scale vortex transfer over the entire thickness of the atmosphere along the meridian is nothing but a small difference of considerably high macroturbulent fluxes toward the poles in the middle and lower stratosphere and toward the equator in the upper and middle troposphere.

The maximum zonal values of the vertical flux of ozone are observed in the zones of downward branches of the mean meridional circulation. However, the global average of the vertical macroturbulent flux of ozone exceeds by several times the vertical flux caused by meridional circulation.

136   The principal shortcoming of the Lawrence Livermore Laboratory model lies in the fact that it gives lower values of total ozone content and temperature in the upper troposphere and stratosphere of polar regions of both the hemisphere (for example, the temperature at the poles is lower by approximately 20 K). Apparently, this is related with the fact that numerical modeling does not ensure the progression of very long wavelengths with higher amplitudes in the troposphere and stratosphere and therefore the macroturbulent transfer of ozone and heat directed toward the poles appears to be underrated. Now the question whether an increase in the amplitude of very long waves and increase of ozone and heat transfer directed toward the poles would occur when the duration in numerical modeling is increased, remains to be discussed. A comparison of calculated and observed mean fields of the zonal components of velocity shows a satisfactory agreement. The main discrepancy lies in the overestimation of two tropospheric maximums of transfer rates from west to east and a similar maximum in the stratosphere of the hemisphere experiencing winter.

The application of modern fast computers enables us to achieve an integration of the complete system of equations of thermo- and hydro dynamics without taking chemical reactions into account (in this, resolutions of 3 km along the vertical and 250 km in horizontal directions are attainable). Further addition of even a minimal set of equations describing the kinetics of reactions (the complete set consists of 100 equations) leads to an extremely large volume of computations. Therefore, a three-dimensional model of the stratosphere developed at the Massachusetts Institute of Technology is based on the application of a geostrophic approximation and a resolution (of the order of 1500 km at temperate latitudes) in the horizontal direction (resolution in the vertical direction for the 0–70 km layer equals 3 km). In order to simplify the problem, only the equations describing the kinetics of ozone have been considered. Thus, the equilibrium distribution of NO_x should be given *a priori*.

The two-dimensional zonal models found wider application for practically important estimates of the impact of ozone on climate. In this case, the interaction of dynamics and chemistry is determined by the set of equations of the kinetics of small components, written in the spherical system of coordinates:

$$\frac{\partial [R_i]}{\partial t} + \frac{[u]}{a}\frac{\partial [R_i]}{\partial t} + [\omega]\frac{\partial [R_i]}{\partial t} + \frac{1}{a\cos\vartheta}\frac{\partial}{\partial\vartheta}\cos\vartheta$$

$$\times \left[\overline{R_i^* u_i^*}\right] + \frac{\partial}{\partial p}\left[\overline{R_i \omega_i^*}\right] = [P_i - L_i], \tag{II.6}$$

where $i = 1, 2, \ldots, n$.

137   Here $R_i = n_i/n$, n_i and n being the concentrations (molecules/cm^3) of the ith component of minor impurities and air, respectively, t the time, a the earth's radius, u the horizontal component of velocity, p the atmospheric pressure, $w = dp/dt$, ϑ the latitude, and P_i, L_i the sources and sinks of minor impurities. The square brackets denote zonal averaging, and the asterisks signify deviation from the zonal average:

$$X^* u^* = -\left[K_{yy}\frac{\partial [\overline{X}]}{\partial y} + K_{yz}\frac{\partial [\overline{X}]}{\partial z}\right];$$

where the coordinate y is measured in the direction of the north.

A similar expression determines the vertical turbulent fluxes. The coefficients of turbulent mixing K_{yy}, K_{yz} and K_{zz} are given from the data of observations. In the system of equations (II.6), the second and third terms on the left characterize the transfer of matter caused by meridional circulation, the fourth term represents the horizontal macroturbulent flux directed toward the pole, and the fifth vertical turbulent mixing.

An independent (from the equations of thermo- and hydro-dynamics) study of this system indicates the impossibility of consideration of feedback links between changes of chemical composition of the atmosphere and the thermodynamic regime, but at the same time enables us to take into account a sufficiently comprehensive set of chemical reactions. This opens up possibilities of assessing the impact of atmospheric pollution in the ozone layer. A detailed discussion on the question of applicability of two-dimensional models can be found in the work of Hidalgo [96].

For an approximate assessment, several variants of the one-dimensional model, based on the application of a globally averaged equation of kinetics of minor components, were worked out. The equation is:

$$\frac{\partial (\rho \tilde{R}_i)}{\partial t} = \frac{\partial}{\partial z}\bar{\rho} K_z \frac{\partial \tilde{R}_i}{\partial z} + \rho \tilde{P}_i - \rho \tilde{L}_i. \tag{II.7}$$

Here the sign of the wave denotes global averaging and ρ the density of air. The use of the MIT two-dimensional model for calculations of an unperturbed field of ozone concentration showed that in order to attain

a state of equilibrium, the computation is to be carried out over two years (in time steps of 10 minutes). In this case, the absolute values of the total ozone content, as well as their space-time variability are simulated very satisfactorily. The chemistry of the perturbed stratosphere is determined by the impact of supersonic aviation on ozone, caused by injection of NO_x. These injections cause a vertical transfer of NO_x into the middle
138 stratosphere (altitude 20–35 km) and a large-scale advection in the lower and middle stratosphere, are propagated on a global scale.

Hunten [98] obtained estimates of the impact of small quantities of nitrogen in the form of oxides (bearing in mind mainly nitrogen dioxide) arising as a result of combustion products from subsonic and supersonic aircraft, on the stratospheric ozone within the framework of an approximate one-dimensional model of turbulent diffusion in the troposphere and stratosphere. The vertical profile of the coefficient of mixing (cm²/s) has been found on the basis of the measurement data of methane concentration at altitudes of up to 50 km and approximated by an analytical relationship $K_z = 2200 \exp [(z-14)/9.43]$ (z being the altitude in kilometers). In the 'tropopause transport' approximation, under consideration, the altitude is measured from the level of the middle tropopause, and K_z is considered as a purely empirical parameter. Calculations resulted in the values presented in Table II.6.

Table II.6. Assessment of impact of subsonic and supersonic aircraft on atmospheric ozone

Parameters	Subsonic aircraft	Supersonic aircraft	
		existing	future
Number of aircraft	1000	1000	60	200
Injection magnitudes, kt/year	600	600	60	800
Altitudes of release, km	12	13	16	20
Concentration of odd nitrogen, ppb	0.34	1.22	0.43	12.3
Relative decrease in ozone content, %	0.47	1.7	0.60	17

The very high sensitivity of the impact of nitrogen oxides on ozone to the altitude of release must be studied in detail.

Analysis of the results of numerical modeling revealed a high sensitivity of the impact of ozone to the selection of the eddy-diffusion coefficient. In order to find ways to reduce the discrepancies of results obtained in [102], an attempt has been made to use in one-dimensional models of the stratosphere the data of aircraft and balloon observations on the global distribution of ^{14}C-isotope concentration, carried out during and after nuclear detonations during the fifties and sixties, of the adopted vertical profiles of the eddy-diffusion coefficient on K_z.

For the period 1955–1967, the regular observational data on the excess
139 (as compared to background) concentration of carbon-14 in the stratosphere and troposphere, averaged over three-month intervals is available.
After completion of the 1962 tests, a cloud of carbon-14 was observed
encompassing the entire northern hemisphere and with a maximum concentration in the 19–20 km layer with an extremely small dispersion along
the vertical, which is very similar to the conditions of stratospheric pollution caused by supersonic aircraft. The data on excess concentration
of carbon-14 is of immense importance for testing the adequacy of two-
and three-dimensional models, since this data appears to be a reliable
indicator of transport parameters in the stratosphere.

In [102], nine models used earlier, of vertical profiles of the eddy-
diffusion coefficient, have been analyzed, and it was shown that those
models are most realistic for which K_z is minimum in the 15–20 km layer
and increases in the 20–50 km altitudinal range.

As has been mentioned earlier, the preliminary assessment of the impact on ozone appeared to be erroneous.

In view of the principal role of ozone in the evolution of the atmosphere and the earth, a calculation of fluctuations of the ozone content in
the geologic past has been carried out [42]. The computations were
worked out taking into account the probable composition and structure
of the paleoatmosphere, which particularly contained a considerably
lower amount of oxygen as compared to the contemporay atmosphere
(present-day oxygen content is taken as unity). A photochemical model
of the atmosphere, applied for calculations of the ozone evolution, had
been developed with the consideration of reactions (in addition to the
'classical' chemistry of ozone), in which water vapor, hydroxyl and
hydropyroxyl radicals, nitrogen oxides, methane and carbon dioxide were
also involved.

The computations, worked out for various models of the atmosphere
have shown that, due to a lack of certainty regarding atmospheric pressure near the earth's surface and concentration of the odd oxides of
nitrogen, there always exists an ozone screen within the limits of a
reasonable assumption regarding parameters of the paleoatmosphere
(even if the oxygen concentration is taken as 10^{-3}). For an ozone content
of 0.3 and an oxygen level at 4×10^{-3} (as opposed to the present-day
values) the total ozone content comprised 50% of the contemporary value.
If abiotic dissociation of water vapor could provide an oxygen content
of only 10^{-3} of the contemporary value, then it follows that ultraviolet
solar radiation was possibly not a sufficient source of energy for a prebiotic synthesis of organic molecules on the earth's surface. In such a
case either the existence of another source of energy, or the possibility of
a prebiotic synthesis beyond the earth's environment should be assumed.

140   The results of recent radioastronomical detection of organic molecules
in the dark section of the nebulae demonstrate the probability of chemical
synthesis beyond the earth. Complex molecules of extraterrestrial origin
could have accumulated on the earth's surface as a result of intense
meteoric showers, during the first eons of the earth's history. Finally, the
results thus obtained show, irrespective of the origin of prebiotic organic
molecules, that, for the most varied assumptions regarding the paleoatmosphere the subsequent biological evolution continued in the presence
of an ozone shield.

Adequacy of averaging and sensitivity to input parameters. The use of
one- or two-dimensional models of the atmosphere for numerical modeling of the stratospheric composition and its anthropogenically caused
fluctuations brings forth the problem of analyzing the adequacy of averaging, related with the application of such models from the viewpoint of
characteristics of zonal and global averages of various parameters, and
the correctness of the description of various relationships. The relationships can be sufficiently reliably revealed only by three-dimensional
modeling. If we talk of a chemical reaction of the type $A+B \rightarrow C+D$,
with a reaction rate k, then it is obvious that, in general $\overline{kAB} \neq \overline{k}\overline{A}\overline{B}$.
However, the absence of sufficiently comprehensive observational data
precludes the possibility of empirically testing such relationships.

Thus in [173] an analysis of the adequacy of averaging the data of
three- and two-dimensional numerical models was undertaken. The three-
dimensional 13-level (0–44 km) model developed by the faculty of the
British Meteorological Service takes into account only the profiles of water
vapor, ozone and oxides of nitrogen NO_x ($NO+O_3 \rightarrow NO_2+O_2$) released by aircraft. Therefore, the actual value $\bar{P}_{act} = \overline{k\,[NO] \cdot [O_3]}$ and the
product of zonally averaged values $\bar{P}_{2D} = \overline{k}\overline{[NO]} \cdot \overline{[O_3]}$ obtained from the
two-dimensional model, taking into account 20 reactions, have been
compared.

A similar comparison has been carried out for results, obtained from
the two- and one-dimensional models (analysis of adequacy of global
averaging). The calculations for January and July carried out on the basis
of two- and one-dimensional models reveal significant discrepancies of
products of the type $\overline{kAB}$ and $\overline{k}\overline{A}\overline{B}$ for various reactions with averaging
both over the hemispheres, as well as over the entire globe. The discrep-
ancies are especially large in the upper troposphere and the lower stratosphere. In the case of reactions such as

$$OH + NO_2\,(+M) \rightarrow HNO_3\,(+M),$$

the deviations from actual average values in the stratosphere attain 200%
141 and then determined almost completely by correlations between [OH]
and [NO₂] but not by the dependence on pressure or temperature. The

one-dimensional model indicates acceptable results only for the middle and upper stratosphere and is valid for summer when photochemical processes dominate.

A similar situation observed in the analysis of a two-dimensional model, does not in the least provide reliable quantitative results for the troposphere, where the quantities $100\,(\bar{P}_{act} - \bar{P}_{2D})/\bar{P}_{act}$ for reactions between NO and O_3 are negative and attain values of 30–35% (in the remaining cases up to 50%). Satisfactory results are obtained only for the middle stratosphere during summer. The effect of the opposite sign (up to 90% in the zone of tropospheric jet flow) causes a covariance of $[\overline{H_2O}]\cdot[O_3]$. The main reason for the discrepancy is the nonlinear correlation between [NO] and $[O_3]$.

It is known that the zonal, hemispheric and global averages can be reliably calculated, either by using the data of local measurements in a real atmospheric layer, or (for an approximate estimation) by computations on the basis of three-dimensional models. From the standpoint of chemical composition, the real atmosphere is, as a rule, inhomogeneous and characterized by nonlinear correlations over a much broader interval of spatial scales than can be obtained with the help of the corresponding numerical models. The latter indicate that the estimates from the three-dimensional models, inevitably ignoring the intervals of smaller scales, would provide smaller deviations from the actual average values.

Another important aspect of theoretical modeling of the stratosphere lies in the estimation of the sensitivity of results obtained to the input parameters and, particularly, to the reaction rates. For numerical experiments, which are aimed at estimating the effect of inaccuracy in assigning the input parameters on the results of the anthropogenic influence on ozone layers, Rundel et al. [150] developed a one-dimensional steady-state (averaged over a day) photochemical model of the stratosphere. In this model a 15–60 km layer of the atmosphere in vertical steps of 5 km each has been examined. The cycles of odd oxygen ($O_x = O + O_3$), odd nitrogen ($NO_x = N + NO + NO_2 + NO_3 + ClONO_2 + HNO_3$), odd hydrogen ($HO_x = H + OH + HO_2, 2H_2O_2, HCl, HNO_3$) and odd chlorine ($ClO_x$), as well as chain reactions of methane oxidation have also been fairly comprehensively considered. The vertical profiles of temperature, nitrogen concentration, oxygen and carbon dioxide are given. The interaction technique enables us to take into account the interaction between various components.

142   The concentrations of odd oxygen, water vapor, H_2, CO, N_2O, odd nitrogen, CH_4, CH_3Cl, CCl_4, CF_2Cl_2, $CFCl_3$ and odd chlorine have been calculated by using equations of diffusion with consideration of corresponding sources and sinks. Concentrations of odd hydrogen, products of methane oxidation and relative concentrations of various components

related to the O_x, NO_x and ClO_x cycles were obtained from the conditions of photochemical equilibrium. The effect of the daily trend of chemical processes, being reflected, for example, in the formation of chlorine nitrate, is taken into account by an analytic approximation. Rundel et al. [150] simulated in detail a model for 55 reactions and coefficients of their reaction rates as well as photodissociation cross reaction of various components depending on the wavelength.

Modifications in reaction rates of $NO + HO_2$ and a few other reactions led to fairly significant changes: strengthening of odd oxygen formation at lower altitudes as a result of $NO + HO_2$ reaction, weakening of the role of $NO_2 + O$ and strengthening of $OH + O_3$ reactions as mechanisms of depletion of O_x, and the increasing contribution of $ClO + O$ in the destruction of odd oxygen at altitudes of 20–25 km. All this attenuated the sensitivity of the model to NO_x, but strengthened it toward ClO_x. A comparison of the results of computations obtained on the basis of a simplified model under consideration, with the data of more comprehensive models and observations, enabled us to reveal a satisfactory correspondence. Hence the possibility of use of a one-dimensional model for the estimation of the effect of inaccurate assumptions regarding input parameters on calculations of the anthropogenic influences on the ozone layer.

Even a relatively simple one-dimensional model of the stratosphere contains many input parameters. Errors in the assumption of these parameters determine the reliability of the end results: reaction rates, absorption and scattering cross sections, intensity of solar radiation, vertical eddy-diffusion coefficient, temperature profiles and composition of the atmosphere (nitrogen and oxygen concentrations). The most important sources of error in the estimation of an effect on the ozone layer are uncertainties in information on reaction rates and vertical eddy diffusion. The latter is usually characterized by assuming a diffusion coefficient, which should approximate the averaged effect of the general global circulation of the atmosphere. Such approximations are not yet defined with sufficient accuracy in order to enable a quantitative assessment of their reliability.

Stolarski et al. [162] have, therefore, studied the effect of errors in assigning input parameters on the accuracy of estimates of the impact on stratospheric ozone, confining themselves only to an analysis of the contribution of errors in reaction rate coefficients. Insofar as the results, 143 obtained on the basis of the cited one-dimensional model, are most sensitive in about 10 out of a total set of 55 considered reactions, calculations using the Monte Carlo method of 'spread of indeterminacies' in the model of the stratosphere, required about 2^{10}, or approximately 1000 cases. This enabled us to carry out computation within a short computer time (computation of 2000 cases on an 'IBM-360-91' requires one hour of computer time) and has provided an excellent agreement in results.

A specific feature of the discussed numerical modeling is the estimation of reliability of the obtained results taking into account the simultaneous variation of various input parameters and interactions between them. The uncertainty in assigning the input parameters varies within wide limits, characterized by the coefficient with values between 1.1 and 10. The probability of error distribution has been approximated by the logarithmic normal law, while the first numerical experiment involves the calculation of vertical profiles of ozone concentration and other components for the 'best set' of reaction rates.

The background concentration of Cl_x has been accepted as 1.5 ppbv which is determined by preassigning the concentration of methyl chloride as 1.0 ppb, carbon tetrachloride as 0.1 ppb and of Cl_x as 0.1 ppb at the lower boundary of the considered layer of the atmosphere. Subsequently, on the background concentration of the chlorine fraction, a perturbation of the prescribed level is superimposed and the contents of ozone and other elements are computed. The results thus obtained are used for comparison with the data of numerical modeling for perturbed values of the rates of all the 55 reactions and concentrations of Cl_x equal to 1.5 ppbv.

The next step in the calculations consists in superimposing a perturbation on the background concentration of the chlorine fraction. The process of calculations continues for various combinations of random perturbations of 55 reaction rates and without perturbations. The final result is expressed in the form of statistical characteristics signifying the variability of the vertical profiles of ozone concentration, as well as the most important components of the Chapman, hydrogen, nitrogen and chlorine cycles.

If the perturbation of odd chlorine concentration is prescribed as 5.7 ppbv at the background value of 1.5 ppbv, then the imprecision caused by this in the estimates of the impact on ozone within the limits of 1 σ is 1.35 for the upper limit of the uncertainty interval and 1.69 for the lower limit. The values 1.81 and 2.86 correspond to the case of 2 σ. For these estimates the assumption is made that releases of Freon-11 and -12 remain at the level of 1975 till a steady state is attained.

Stolarski et al. [162] simulated, in detail, data on the limits of variations in the vertical profiles of ozone and other components, caused by the uncertainty of data on reaction rates. It has been shown that errors in the estimation of concentration of odd oxygen at altitudes of 20 km or more (below 50 km it is purely ozone) are caused exceptionally by the reactions of ozone destruction, and within the limits of the entire stratosphere the reaction $NO_2 + O \rightarrow NO + O_2$ dominates.

When the predominant form of the chlorine fraction is HCl, the error

136

in the estimates of chlorine concentration is very small. In the case of components of the nitrogen cycle, errors are not large but may go up due to an uncertainty of the characteristics of the photolytic process, which happens to be the main mechanism of NO_2 and HNO_3 destruction at almost all altitudes. Calculations carried out earlier, led to the conclusion that errors in assigning the extraterrestrial solar radiation and cross sections of photolysis have little overall effect on the general impact of the chlorine cycle on ozone.

Verification of the obtained results by comparing them with the observational data is meaningful only where there is sufficiently reliable and comprehensive data for the characterization of mean values of concentrations of various components at temperate latitudes and in the global stratosphere. Presently, such a situation is indicated only for ozone. Calculations of ozone concentration at the tropopause level give values which are considerably underestimated as compared to those observed, whereas a variation of reaction rates within various limits does not provide agreement between calculated and measured values. With the accumulation of measurement data on the composition of the stratosphere (primarily components not having diurnal variations), there will be a greater possibility of comparison of the results obtained by numerical modeling with experimental values.

Butler [47] accomplished numerical experimentation on the sensitivity of a model in such a way that at first the calculation is carried out till a state of equilibrium is obtained for the given set of parameters and the calculation is then repeated with selected changed parameters (reaction rate, boundary conditions, solar radiation flux, photolytic cross section) within the limits of $1\,\sigma$ or $2\,\sigma$. If R is the concentration of the required component (to be calculated) and P a variable parameter, then in place of the sensitivity characteristic, the specific uncertainty $U_c = \log P_{+1\sigma} - \log P_{-1\sigma}$ is used and the sensitivity is given by $S_c = U_{+1\sigma}/(\log R_{+1\sigma} - \log R_{-1\sigma})$. Similarly $U_{+2\sigma}$ and $U_{-2\sigma}$ as also the corresponding sensitivities can be determined. The total uncertainty of the result T is determined from $T = \exp\left(\sqrt{\sum_i U_i^2}\right)$.

Analysis of calculations of U_c and S_c for various factors, determining the concentration of ozone (at altitudes of 30 and 50 km), nitrogen fraction (35 km), methane (40 km), nitrous oxide (30 km), ratio of concentrations Cl/ClO (35 km and 40 km) demonstrates that:

1) the adequacy of contemporary models of the stratosphere is doubtful (possible uncertainty of results is very high);

2) the vertical profile of the coefficient of turbulent mixing, used for calculations, is unreliable; and

3) testing of models by way of comparison, with only two components with measured concentrations, is not sufficiently reliable.

The fact, that sources of errors depend considerably on altitude is a very important factor.

Butler and Stolarski [48] carried out calculations of the combined effect on the composition of the stratosphere (concentrations of O_3, CH_4 and CH_3Cl) of an increase in concentration of such pollutants as CO, NO_x and chlorofluoromethanes, on the basis of application of a steady one-dimensional model of the atmosphere. The variation with time has been obtained by calculations carried out at intervals over a 20-year period. It is assumed that the globally averaged contribution of anthropogenic carbon monoxide comprises 25% of a net flux of carbon monoxide, equal to 3.9×10^{11} molecules/cm^2 s. The anthropogenic production of oxides of nitrogen equals 33% for a net flux of 2.4×10^{10} molecules/cm^2 s. It is assumed that an increase in the concentration of CO and NO_x constitutes 2 or 4% per year. A rise in the chlorofluoromethane concentration is measured against a background of constant releases. The calculations of fluctuations in the concentrations of minor components under consideration cover the period 1930–2030.

The results of computations carried out for the dynamics of the total ozone content show that it roughly increased up to 1970, and thereafter decreased fairly rapidly as a result of the increasing effect of chlorofluorocarbons. The long-term trend of ozone closely depends on the variations in the ratios of concentration of the various pollutants. It is assumed, for example, that a rise in CO and NO_x concentrations for a constant release of chlorofluoromethanes constitutes 4%; therefore the effect of carbon monoxide and nitrogen oxides on the total ozone-content increase will start dominating after the year 2000. For a 2% increase, the effect of chlorofluoromethanes will remain predominant till the year 2030. The calculated increase of ozone content during 1960–1970 is not more than 0.1%, and the decrease after 1970 also does not exceed 0.5%, i.e. it is impossible to detect these trends from observational data. The results obtained indicate that the impact of various anthropogenic factors on ozone seriously complicates the assessment of trends caused by chlorofluoromethanes in the past and in the future.

As already mentioned, while summing up the accomplishment of CIAP in 1975, it was concluded that a fleet of 120 supersonic *Concorde*-type airplanes should cause a drop in the total ozone content in the range of 0.5 to 0.9%. Subsequent calculations, results of which were published in 1977, led to the conclusion that there should be an increase, rather than a decrease, in the ozone content. Such a radical evolution in the releases, obtained on the basis of numerical modeling, is related mainly with the gradual modification regarding the contributions of various chemical processes in changing the ozone concentration. In particular, it refers to the reactions of the chlorine cycle and the data on the reaction rates.

Smith [158] studied the possibility of a statistical approach to the estimation of the confidence interval for the decrease in ozone content, determined on the basis of uncertainty of reaction rates, which are given by the logarithmic normal distribution. A comparison of the data on the rates of 46 reactions, available in 1975 and 1977, revealed a tendency toward the overrating of 'central' values of reaction rates or underestimation of errors in the rates of slow reactions (the latter may be related to the fact that measurements of rates of slow reactions are less accurate). Here, it is obvious that the data available in 1975 on the reaction rates was not obtained with a confidence level of 95% in the limits of $\pm 2\sigma$: this data appeared to be full of systematic errors which could not be discovered at that time. There is also no basis for considering that the reliability of data for the year 1977 has been established in the limits of $\pm 2\sigma$.

An earlier analysis led to the conclusion that there is a considerable decrease in photochemically caused increases in the rates of thermal relaxation and photochemical relaxation of ozone in the upper stratosphere, under the influence of variations in the transparency of the atmosphere. In the case of strong fluctuations of the vertical profile of ozone concentrations, variations in the rate of ozone dissociation and warming were observed as a result of considerable changes in transparency. Such a situation demonstrates that there may exist a nonequilibrium concentration of ozone in the stratosphere. Strobel [163] examined some numerical examples illustrating this phenomenon, and showed that instability in the ozone profiles emerges only in the case of geophysically nonfeasible interactions. Such interactions indicate transient gains, 147 which are followed by a gradual damping, leading to an unperturbed state. The analytical formulas, describing thermal and photochemical relaxations in the case of such perturbations, the altitudinal scale of which is comparable with that for ozone, have been obtained.

3.2. Stratospheric reactions of formation and destruction of ozone. In accordance with the existing concepts [57, 95, 96], formation of ozone takes place as a result of the Chapman cycle of photochemical disintegration of molecular oxygen by ultravoilet solar radiation in the upper stratosphere with the following combination of reagents:

$$O_2 + h\nu \rightarrow O + O \ (\lambda < 242 \text{ nm}); \tag{II.8}$$

$$O + O_2 + M \rightarrow O_3 + M; \quad k_2 = 1.1 \times 10^{-34} \exp(500/T); \tag{II.9}$$

(M is some other gas; k_i indicates the reaction rate).

These processes exist in dynamic equilibrium with the natural processes of photolytic destruction of ozone:

$$O_3 + h\nu \rightarrow O_2 + O(^1D) \quad (\lambda < 310 \text{ nm}); \tag{II.9a}$$

$$O_3 + h\nu \rightarrow O_2 + O \quad (\lambda < 1140 \text{ nm}); \qquad \text{(II.9b)}$$

as well as chemical bonding between ozone and atomic oxygen.

The second form of interaction, primarily, includes a reaction with the participation of atomic oxygen:

$$O_3 + O \rightarrow O_2 + O_2; \quad k_4 = 1.9 \times 10^{-11} \exp(-2300/T). \qquad \text{(II.10)}$$

The fifth reaction of the Chapman cycle is as follows:

$$O(^1D) + M \rightarrow O + M; \quad k_5 = 6 \times 10^{-11}. \qquad \text{(II.11)}$$

The rate coefficients are expressed here either in $cm^6/molecules^2$ s (k_2 is the three-body problem) or in $cm^3/molecules$ s (reaction with participation of two components).

Reactions of the hydrogen cycle HO_x play an important role in the formation of ozone. A part of these reactions determines the relative concentrations of the components associated with the hydrogen cycle:

$$O + OH \rightarrow H + O_2, \quad k_6 = 4.2 \times 10^{-11}; \qquad \text{(II.12)}$$

$$O_3 + OH \rightarrow HO_2 + O_2, \quad k_7 = 1.6 \times 10^{-12} \exp(-1000/T); \qquad \text{(II.13)}$$

$$CO + OH \rightarrow H + CO_2, \quad k_8 = 2.1 \times 10^{-13} \exp(-75/T); \qquad \text{(II.14)}$$

148

$$H_2 + OH \rightarrow H + H_2O, \quad k_9 = 2.3 \times 10^{-11} \exp(-2450/T); \qquad \text{(II.15)}$$

$$O_3 + H \rightarrow OH + O_2, \quad k_{10} = 2.6 \times 10^{-11}; \qquad \text{(II.16)}$$

$$O_2 + H + M \rightarrow HO_2 + M, \quad k_{11} = 2.1 \times 10^{-32} \exp(290/T); \qquad \text{(II.17)}$$

$$O + HO_2 \rightarrow OH + O_2, \quad k_{12} = 2.0 \times 10^{-11}; \qquad \text{(II.18)}$$

$$O_3 + HO_2 \rightarrow OH + 2O_2, \quad k_{13} = 1.0 \times 10^{-13} \exp(-1250/T); \qquad \text{(II.19)}$$

$$HO_2 + NO \rightarrow OH + NO_2, \quad k_{14} = 2.2 \times 10^{-13}; \qquad \text{(II.20)}$$

$$OH + NO_2 (+M) \rightarrow HNO_3 (+M), \quad \left.\begin{array}{l} k_{15} = [4.2 \times 10^{-22} \exp \\ \qquad \times (-170/T) (M)]\sigma^{-1}; \\ \sigma = 4 \times 10^{-11} (M) \\ \qquad + 1.6 \times 10^{-2}T; \end{array}\right. \qquad \text{(II.21)}$$

$$HNO_3 + h\nu \rightarrow OH + NO_2, \quad \left.\begin{array}{l} \\ \lambda < 546 \text{ nm} \end{array}\right\} \qquad \text{(II.22)}$$

$$HO_2 + HO_2 \rightarrow H_2O_2 + O_2, \quad k_{17} = 3.0 \times 10^{-11} \exp(-500/T); \qquad \text{(II.23)}$$

$$H_2O_2 + h\nu \rightarrow OH + OH, \quad \lambda < 565 \text{ nm}. \qquad \text{(II.24)}$$

It should be noted that recent direct measurements gave the following value:

$$k_{13} = (1.4 \pm 0.4) \times 10^{-14} \exp[(-580 \pm 100)/T],$$

but a much lower energy of activation, that accelerates the reaction rate approximately 6 times the earlier value (in the conditions of the stratosphere). An increased efficiency of hydroxyl formation can signify a stronger ClO_x impact (through an increase in the rate of $OH + HCl$ interaction—see further) and weaker NO_x impact (as a result of increase in the rate of $OH + NO_2 + M \rightarrow HNO_3 + M$ reaction) on the ozone layer.

Another group of reactions describes the sources and sinks of the hydrogen fraction:

$$H_2O + O(^1D) \rightarrow OH + OH, \quad k_{19a} = 3.0 \times 10^{-10}; \quad \text{(II.25a)}$$

$$CH_4 + O(^1D) \rightarrow CH_3 + OH, \quad k_{19b} = 3.0 \times 10^{-10} \quad \text{(II.25b)}$$

$$H_2 + O(^1D) \rightarrow H + OH, \quad k_{19c} = 3.0 \times 10^{-10}; \quad \text{(II.25c)}$$

$$H + HO_2 \rightarrow H_2 + O_2, \quad k_{20} = 10^{-11}; \quad \text{(II.26)}$$

$$OH + HO_2 \rightarrow H_2O + O_2, \quad k_{21} = 5.0 \times 10^{-11}; \quad \text{(II.27)}$$

$$OH + H_2O_2 \rightarrow H_2O + HO_2, \quad k_{22} = 1.7 \times 10^{-11} \exp(-900/T); \text{(II.28)}$$

$$OH + HNO_3 \rightarrow H_2O + NO_3, \quad k_{23} = 9 \times 10^{-14}. \quad \text{(II.29)}$$

As Hidalgo and Crutzen [95] have shown, reactions of the cycle, related with the oxidation of methane and responsible for the formation of ozone in the troposphere and lower stratosphere, are of great importance:

$$OH + CH_4 \rightarrow H_2O + CH_3, \quad k_{24} = 2.5 \times 10^{-12} \exp(-1660/T); \text{(II.30)}$$

$$CH_3 + O_2 + M \rightarrow CH_3O_2 + M, \quad k_{25} = 2.6 \times 10^{-31}; \quad \text{(II.31)}$$

$$CH_3O_2 + NO \rightarrow CH_3O + NO_2, \quad k_{26} = 1.5 \times 10^{-12} \exp(-500/T); \quad \text{(II.32a)}$$

$$CH_3O_2 + HO_2 \rightarrow CH_3O_2H + O_2, \quad k_{27} = 3.0 \times 10^{-11} \exp(-500/T); \quad \text{(II.32b)}$$

$$CH_3O_2H + h\nu \rightarrow CH_3O + OH; \quad \text{(II.33)}$$

$$CH_3O + O_2 \rightarrow CH_2O + HO_2, \quad k_{29} = 4.2 \times 10^{-13} \exp(-3000/T); \quad \text{(II.34)}$$

$$CH_2O + h\nu \rightarrow H_2 + CO, \quad \lambda \leqslant 350 \text{ nm}, \quad J_{30a} = 1.1 \times 10^{-4}; \text{(II.35a)}$$

$$CH_2O + h\nu \rightarrow H + CHO, \quad \lambda \lesssim 350 \text{ nm}, \quad J_{30b} = 3.3 \times 10^{-5}; \text{(II.35b)}$$

$$CH_2O + OH \rightarrow H_2O + CHO, \quad k_{31} = 1.4 \times 10^{-11}; \quad \text{(II.36)}$$

$$CHO + O_2 \rightarrow HO_2 + CO; \quad k_{32} = 1.7 \times 10^{-13}. \quad \text{(II.37)}$$

Insofar as methane cannot be formed in the stratosphere, its concentration, while decreasing above the tropopause is determined by the rates of chemical reactions leading to the destruction of methane and by the vertical eddy-diffusion coefficient. This last condition impelled many authors to use data on vertical profiles of methane concentration for estimating the parameters characterizing vertical transfer. It is natural that in such a case reliable data on methane concentration at different altitudes must be available.

In view of this Ackerman et al. [35] undertook the task of analyzing the results obtained earlier, and of fresh measurements of methane concentration in the 22–35 km layer with the help of high-resolution infrared spectroscopy. This enabled us (taking into account fresh data on time-structure parameters of the absorption spectrum) to incorporate certain modifications in the earlier results and eliminate existing contradictions. Information on vertical profiles of methane concentration has been obtained from a balloon fitted with a spectrometer provided with a diffraction grating on October 2, 1975, from a polygon of the French Space Research Center.

Measurements at 43°N and 2°E were conducted for the altitudinal variation of the sum from 80.5 to 95.5°. The concentration of methane above the sounding ceiling of 75 km was obtained on the assumption that 150 the atmosphere remains homogeneous to a height of 6 km, and the concentration of methane molecules at the altitude of 75 km is 4.5×10^{10} cm^{-3}. An account of all the measurement data, obtained over 10 years by different authors, enabled us to obtain the methane concentration for the 30 km altitude (Table II.7).

Table II.7. Concentration of methane at 30 km altitude

Concentration, 10^{-11} mol/cm^3	Date of measurement	Latitude
1.9	August, 1965	47°N
2	December, 1967	33
2	October, 1971	43
2.2	October, 1975	43
3	—	32

If all possible errors in the measurements are taken into account, then according to data given in Table II.7, a conclusion regarding the high stability of methane concentration at an altitude of 30 km, can be drawn irrespective of the latitude and time of the year. In [35], on the basis of all available data the most probable vertical profile of methane concentration and the corresponding vertical profile of the coefficient of turbulent mixing, has been obtained. In the 20–40 km layer the ratio of methane mixture decreases approximately from 1.0 to 0.1 ppm. The low

142

methane concentration in the stratosphere is of great importance for model calculations of the stratospheric composition and, particularly, leads to a decreasing ratio of concentration, namely HCl/ClO.

Investigations carried out earlier showed that in many processes of self-cleaning, taking place in the troposphere, the hydroxyl ions play a highly important role. It controls the concentration of numerous weakly soluble minor gaseous components, such as methane, carbon monoxide, hydrogen sulfide, sulfur dioxide, CH_3Cl, $CH_xCl_yF_z$ and CH_xBr_y. In [167] the effect of increasing the anthropogenic release of carbon monoxide on the tropospheric OH has been analyzed and the subsequent evolution of the $CO-OH-CH_4$ cycle studied. It has been proposed that the chief mechanism in the discharge of methane and carbon dioxide is the following reaction:

$$OH + CO \rightarrow CO_2 + H. \tag{II.30a}$$

Subsequently, oxidation of CH_3 serves as the main source for the formation of natural carbon monoxide.

151 These reactions form the main mechanism of production of tropospheric OH. Anthropogenic generation of carbon monoxide caused by burning of fuel constitutes about 4×10^8 t/year, as compared with natural production of the order of $(1-3) \times 10^9$ t/year (mainly on account of oxidation of ammonia, following reaction II.30). A further increase in global carbon monoxide concentration may lead to a significant decrease in OH concentration. In [167], the concentration of HO_x, CO, NO_x and CH_4 on the basis of application of a one-dimensional model of nonsteady diffusion, with the consideration of three possible alternatives of carbon monoxide concentration dynamics in the future, have been calculated. These computations show that as a result of a decrease in the OH concentration caused by the production of carbon monoxide, a considerable increase in the global concentration of CO, CH_4, CH_3Cl and other minor gaseous components can take place.

At present the level of global carbon monoxide concentration exceeds the level of the fifties by 20-30% and in the case of methane by 10-20%. Even if the production of carbon monoxide is maintained at the present level, the increase in carbon monoxide and methane would be 50 and 25%, respectively, by the year 2025. The time constant characterizing the perturbations of the $CO-OH-CH_4$ cycle, is of the order of several decades.

The variability of this cycle can indirectly affect the chemistry of the stratosphere, as an increase in the methane content can lead to: 1) an increase in the concentration of water vapor and hydrogen fraction (H, OH, HO_2) in the stratosphere, and 2) bonding of a large amount of chlorine following the reaction $Cl + CH_4 \rightarrow HCl + CH_3$, which should promote an increase in the ozone content. Any increase in water vapor and methane would further reinforce the greenhouse effect. The main conclusion states that the contemporary level of anthropogenic release of carbon monoxide is so significant, that it may lead to considerable perturbations in the $CO-OH-CH_4$ cycle.

Whereas the photochemistry of methane in the background conditions of the troposphere and stratosphere has been studied fairly thoroughly, the non-methane-type hydrocarbon compounds (NMHC) in the atmosphere have remained practically unstudied. Meanwhile, for example, such highly active NMHC, as propylene, apparently play a key role in photochemical processes involved in the formation of smog in urban areas. Insofar as the photochemical life of such NMHC is relatively short (in the case of propylene it is about 0.5 day in the troposphere), their concentration in the air is low, which determines the localized nature of the effects of NMHC. A similar situation exists in relation to terpenes produced 152 from vegetation, which exert a significant, though only localized, effect on the formation of ozone and a blue haze in the vicinity of large forests.

An altogether different situation holds for chemically less-active NMHC such as ethane and acetylene, which can attain noticeable concentrations in the troposphere and lower stratosphere. The mean life of C_2H_6 and C_2H_2 in the lower troposphere is about 25 to 50 days respectively. On the basis of application of the one-dimensional model, taking into account the photochemical processes (26 reactions) and vertical turbulent diffusion, Chameides and Cicerone [50] calculated the vertical profiles of concentrations relating to C_2H_6, C_2H_2, C_3H_8, C_4H_{10} and C_5H_{12} in the 0–40 km layer for conditions corresponding to the equilibrium period at 30°N. It is assumed that the earth's surface is the only source of halocarbon compounds. For ethane and acetylene the ratio of the mixture is taken as 1 ppb and for other NMHC it is only 0.1 ppb.

Calculations showed that there may be noticeable quantities of ethane and acetylene in the stratosphere. The main sink of NMHC in the troposphere is determined by reactions with the hydroxyl group. For acetylene this is true in the stratosphere as well, whereas the main sinks of other NMHC in the stratosphere are reactions with chlorine when its concentration does not exceed 10^5 cm^{-3}.

The assessment of the effect of NMHC under consideration, on the cycles of the hydrogen and oxygen fractions as well as chlorine and oxygen in the troposphere and stratosphere, enabled us to draw conclusions regarding the weak influence of NMHC on the background photochemistry of the atmosphere, though it may be mentioned, that reactions which take NMHC into account may be an important source of carbon monoxide.

The global production of carbon monoxide due to oxidation of

144

NMHC can attain values of 220 megatons per year. If in the lower strato-sphere the molecules of vinyl chloride ($CHCl = CH$) appear to be stable, in such a case acetylene is capable of partly averting the destruction of stratospheric ozone caused by Cl and ClO. For low concentrations of ClX in the stratosphere (0.1 ppb or less at a level of 40 km), the NMHC can have a small though significant effect on the chlorine cycle near 30 km. For a more reliable assessment of the role of NMHC, comprehensive measurements of their concentration and modification of the rates of a series of photochemical reactions are essential.

In the beginning of the seventies, the following mechanism of impact on ozone with the participation of nitrogen oxides NO_x was proposed [57]:

$$NO + O_3 \rightarrow NO_2 + O_2, \quad k_{33} = 9.0 \times 10^{-13} \exp(-1200/T); \quad \text{(II.38)}$$

$$NO_2 + O \rightarrow NO + O_2, \quad k_{34} = 9.2 \times 10^{-12}. \quad \text{(II.39)}$$

153   Investigations showed that reactions (II.38) and (II.39) contribute to larger productions than the reactions (II.9), (II.10) and those of the hydrogen cycle, since they proceed more rapidly, whereas nitrogen oxides regenerate from cycle to cycle, thus assuming the role of a catalyst. As it is known, this mechanism created an extremely wide interest not only from the standpoint of its fundamental importance for the development of a general scheme of an ozone sink, but also in connection with the study of the impact of testing of nuclear devices, regular flights of high-altitude supersonic aircraft and launching of powerful transport rockets.

A complete set of reactions of the NO_x cycle includes the following reactions as well [95]:

$$NO_2 + h\nu \rightarrow NO + O, \quad \lambda < 400 \text{ nm}; \quad \text{(II.40)}$$

$$N + O_3 \rightarrow NO + O_2, \quad k_{36} = 3.0 \times 10^{-11} \exp(-1200/T); \quad \text{(II.41)}$$

$$N + O_2 \rightarrow NO + O, \quad k_{37} = 1.1 \times 10^{-14} \exp(-3150/T); \quad \text{(II.42)}$$

$$N + OH \rightarrow NO + H, \quad k_{38} = 5.3 \times 10^{-11}; \quad \text{(II.43)}$$

$$N + HO_2 \rightarrow NO + OH, \quad k_{39} = 2.0 \times 10^{-10}; \quad \text{(II.44)}$$

$$NO + h\nu \rightarrow N + O, \quad \lambda < 191 \text{ nm}; \quad \text{(II.45)}$$

$$N + NO \rightarrow N_2O, \quad k_{41} = 2.7 \times 10^{-11}; \quad \text{(II.46)}$$

$$N_2O + h\nu \rightarrow N_2 + O, \quad \lambda < 337 \text{ nm} \quad \text{(II.47)}$$

$$N_2O + O(^1D) \rightarrow NO + NO, \quad k_{43a} = 1.1 \times 10^{-10}; \quad \text{(II.48a)}$$

$$N_2O + O(^1D) \rightarrow N_2 + O_2, \quad k_{43b} = 1.1 \times 10^{-10}; \quad \text{(II.48b)}$$

$$NO_2 + O_3 \rightarrow NO_3 + O_2, \quad k_{44} = 1.23 \times 10^{-13} \exp(-2470/T); \quad \text{(II.49)}$$

$$NO_3 + h\nu \rightarrow NO_2 + O; \tag{II.50}$$

$$NO_3 + NO \rightarrow 2NO_2, \quad k_{46} = 8.7 \times 10^{-12}; \tag{II.51}$$

$$NO_3 + NO_2 + M \rightarrow N_2O_5 + M, \quad k_{47} = [1.2 \times 10^{-21}\,(M)]/[2 \times 10^8$$
$$+ 1.7 \times 10^{-10}\,(M)]; \tag{II.52}$$

$$N_2O_5 + M \rightarrow NO_2 + NO_3 + M, \quad k_{48} = [4000\,(M)$$
$$\times \exp(-9650/T)]/[2 \times 10^8 + 1.7 \times 10^{-10}\,(M)]; \tag{II.53}$$

$$N_2O_5 + h\nu \rightarrow NO_2 + NO_3. \tag{II.54}$$

Reactions of the methane oxidation cycle H_2O–CH_4–O_x–NO_x determine the balance equation of ozone formation:

$$CH_4 + 4O_2 + 2h\nu_1 + h\nu_2 \rightarrow H_2O + CO + H_2 + 2O_3, \tag{II.55}$$

and consideration of the part of reactions belonging to the oxygen, hydrogen and nitrogen cycles leads to the following balance equation:

$$CO_2 + 2O_2 + h\nu \rightarrow CO_2 + O_3. \tag{II.56}$$

154 When combined, Eqs. (II.55) and (II.56) give

$$CH_4 + 6O_2 \rightarrow H_2O + H_2 + CO_2 + 3O_3. \tag{II.57}$$

On the other hand, if Eq. (II.53a) is replaced by (II.53b), we arrive at a maximum production of ozone on account of methane oxidation:

$$CH_4 + 8O_2 \rightarrow 2H_2O + CO_2 + 4O_3. \tag{II.58}$$

Calculations carried out by Johnston [100, 101] led to the conclusion that only the Chapman (oxygen) reactions and atmospheric circulation maintain about 20% of the balanced production of ozone at altitudes of less than 45 km, and the atmospheric circulation causes intense redistribution of ozone: over a period of a few months, considerable meridional transfer of ozone to the lower stratosphere takes place.

When Chapman reactions alone take place, the global ozone content should increase to twice its value within two weeks. From this follows the necessity of the existence of strong chemical or photochemical reactions destroying the natural ozone. The correctness of this conclusion is confirmed by calculations carried out with different values of the constants determining reactions of the oxygen cycle. Further consideration of reactions related with water vapor (with the participation of H, HO and HOO) shows that they can only explain the destruction of ozone, amounting to 10% of the total value. If the values of nitrogen-dioxide concentrations, obtained from data of recent observations at altitudes up to 36 km, and the corresponding reactions of the ozone cycle are taken into consideration, this ensures coupling of the balance equations since the aforementioned reactions result in the destruction of about 70% of the ozone. As stated earlier, reconsideration of data on the nitrogen cycle led to a conclusion regarding the importance of methane oxidation and the chlorine cycle.

According to Graham and Johnston [85], the interaction between nitrogen oxides of a higher order and ozone is a classical case of such a situation, where complicated reactions, observed in laboratory conditions, can be explained with the help of a series of elementary reactions. On the other hand, the rate coefficients of elementary reactions can be determined by combining the data for several complicated systems. The complicated reactions include: 1) thermal decomposition of nitrogen pentoxide: $2N_2O_5 \rightarrow 4NO_2 + O_2$; 2) interaction of nitrogen pentoxide and nitrous oxide: $N_2O_5 + NO \rightarrow 3NO_2$; 3) formation of nitrogen pentoxide from nitrogen dioxide and ozone: $2NO_2 + O_3 \rightarrow N_2O_5 + O_2$; and 4) decomposition of ozone catalyzed by nitrogen pentoxide: $2O_3 + N_2O_5 \rightarrow 3O_2 + N_2O_5$. These four complex reactions can be quantitatively explained as a result of six elementary chemical reactions with the participation of the free nitrate radical.

155   The photochemistry of the nitrate radical can be of great importance for the ozone budget in the troposphere and lower stratosphere. If the products of nitrate photolysis are nitric oxide and oxygen then the following set of reactions takes place:

$$NO_3 + h\nu \rightarrow NO + O_2;$$
$$NO + O_3 \rightarrow NO_2 + O_2;$$
$$NO_2 + O_3 \rightarrow NO_3 + O_2.$$

The final result of these reactions is the catalytic destruction of ozone:

$$2O_3 + h\nu \rightarrow 3O_2.$$

When the products of photolysis are nitrogen dioxide and atomic oxygen, the following reactions prevail:

$$NO_3 + h\nu \rightarrow NO_2 + O;$$
$$O + O_2 + M \rightarrow O_3 + M;$$
$$NO_2 + O_3 \rightarrow NO_3 + O_2.$$

This set of reactions does not, ultimately, cause any variation in ozone. In view of the aforementioned problems of ozone photochemistry, in [85], the investigation with a view to obtain data on the quantum yield of the nitrate photolytic reaction has been undertaken. A part of the solution to this problem provided information on the optical absorption spectra of a number of oxides and nitrogen oxyacids as well as rates of certain elementary reactions.

Although it is assumed that the kinetics of the N_2O_5–O_3 system is sufficiently well understood, the necessity of further investigations in this area is evident. Therefore, the authors of [85] carried out studies on the kinetics of ozone decomposition catalyzed by nitrogen pentoxide in the dark, and also in the presence of light with photolytic reaction, and is absorbed by the NO_3 radical. They measured the cross section of N_2O_5, HNO_3, NO_2 and NO_3 absorption in ultraviolet, visible and infrared regions of the spectrum. It has been found that the equilibrium constant for the reaction

$$N_2O_5 \rightleftarrows NO_2 + NO_3,$$

in the temperature interval of 298–329 K is $k = (8.4 \pm 1.8) \times 10^{26}$ exp $[-(11180 \pm 100)/T]$ mol/cm^3. The rate coefficients g, m and n (cm^3/mol s) of the following reactions have been measured:

$$2NO_3 \rightarrow 2NO_2 + O_2 \ (g = (8.5 \pm 2.8) \times 10^{-13} \exp[-(2450 \pm 100)/T]);$$

$$O + N_2O_5 \rightarrow \text{products of reaction} \ (m \leqslant 2 \times 10^{-14});$$

$$O + NO_3 \rightarrow O_2 + NO_2 \quad (n = (1.0 \pm 0.4) \times 10^{-11}).$$

156   The rate constants e and f for the following reactions have been found by combining the equilibrium constant K with known, from earlier obtained results, values of K_e and K_f at a temperature of 297 K:

$$NO_2 + NO_3 \rightarrow NO + O_2 + NO_2,$$

$$e = (2.5 \pm 0.5) \times 10^{-14} \exp[-(1230 \pm 100)/T];$$

$$NO + NO_3 \rightarrow 2NO_2; \quad f = (1.9 \pm 0.4) \times 10^{-11}.$$

At the atmospheric pressure of 1013.3 mb, the quantum yield of the nitrate photolytic reaction in the red region of the spectrum is less than unity. In the conditions of the troposphere, with the sun at the zenith, the following values for rate constants of dissociation by solar radiation have been obtained (at a temperature of 298 K):

$$NO_3 + h\nu \rightarrow NO + O_2, \quad j_1 = 0.040 \pm 0.02 \ S^{-1};$$

$$NO_3 + h\nu \rightarrow NO_2 + O, \quad j_2 = 0.099 \pm 0.02 \ S^{-1}.$$

The mean quantum yield for the second process is approximately 0.77 for light in the 470–610 nm wave band, whereas for the first process, it is about 0.23 (in the same wave band), however, in the region of intense absorption in the 610–670 nm wave band it falls to 0.07. The values of the photolytic constant may increase provided the total pressure decreases.

148

The considerable underestimation in earlier values of HNO_3 absorption in the ultraviolet region of the spectrum has been indicated. Absorption in the infrared region of the spectrum attributed to the nitrate radical is, in reality, caused by N_2O_5. The cross section of nitrate absorption in the visible region of the spectrum appeared to be reduced to approximately one-fourth of its value. The cross section of N_2O_5 absorption in the ultraviolet region of the spectrum has been somewhat modified.

Adequate information on the rates of photochemical reactions, responsible for the ozone layer formation, is of exceptional importance. In this connection, Duewer et al. [130] carried out calculations to estimate the sensitivity of catalytic destruction of ozone by nitrogen oxides toward the selection of rates of photochemical reactions determining the process of destruction. The model used in the study of the vertical transfer with consideration of photochemical reactions takes into account 44 reactions of the HO_x, NO_x and O_x cycles with the participation of O, O_3, NO, NO_2, HNO_3, HO, HO_2, H_2O_2 and N_2O for given vertical profiles of water vapor and methane concentration, as well as for the application of a quasi-steady approximation of $O(^1D)$, N and H. Computations have been carried out considering the given combinations of reaction rates for a natural and polluted stratosphere for release of nitrogen
157 oxides at the level of 17 (NO_x-17), 20 (NO_x-20) and 35 (NO_x-35) km (it has been assumed, that over a period of 300 years, 2000 mol/cm² s are introduced into a layer of 1 km thickness), as well as for an increase in water-vapor content in the stratosphere by 10%.

The results of computations characterizing the relative variations of ozone content in the case of four out of five of the aforementioned combinations of reaction rates are given in Table II.8 (k_{21} is the rate coefficient of reaction $HO_2 + HO \rightarrow H_2O + O_2$). As can be seen, for a minimum reaction rate of the stratosphere with release of nitrogen oxide at altitudes of 17 and 20 km, an increase rather than a decrease in total ozone content is observed. In this case, water-vapor concentration in the stratosphere and the following reactions of the HO_x cycle exert a minimum influence on the ozone content in the case of an impact of nitrogen oxides:

$$HO_2 + HO \rightarrow H_2O + O_2$$

(reaction rate in this case is determined by coefficient k_{21}):

$$HO_2 + O_3 \rightarrow HO + 2O_2 \quad (k_{13});$$
$$HO_2 + HO_2 \rightarrow H_2O_2 + O_2 \quad (k_{17});$$
$$HO_2 + NO \rightarrow NO_2 + HO \quad (k_{14});$$
$$HO + NO_2 + M \rightarrow HNO_3 + M \quad (k_{15}).$$

Table II.8. Relative (%) variation of ozone content, caused by releases of nitrogen oxides and decrease of water vapor, for four combinations of reaction rates (A, A', B, C)

Case	A	A'	B	C
NO_x-17	−4.34	−1.75	−8.06	+1.74
NO_x-20		−5.21	−15.92	+1.61
NO_x-35		−14.06	−25.11	−6.42
$-1.1 \times H_2O$	+0.03	−0.19	+0.15	−0.52
$-1.1 \times H_2O$	−4.25	−1.88	−7.89	+1.23

Note: A—CIAP, maximum value; A'—most probable value; B—maximum reaction for release; C—minimum reaction for release.

The last two reactions (k_{14}, k_{15}) are especially important. Even one rate coefficient out of the above-cited reactions (except probably k_{15}) cannot be considered as reliably determined. Where the concentration of water vapor in the stratosphere is considered, it is extremely variable and can be estimated with an accuracy up to a factor exceeding two in the middle and upper stratosphere. All this implies that the uncertainty in assessing the impact of supersonic aviation on ozone is even more serious than was visualized at the time of carrying out the CIAP experiments. The only way to increase the reliability of such estimates appears to be accurate laboratory measurements of reaction rates and more reliable and comprehensive measurement of concentrations of minor components in the stratosphere. The value of k_{21} which can be considered as most probable, gives the reduced ozone content for a fixed release of NO_x. This value of ozone content proves to be half of that predicted in the CIAP report.

Within the dispersion limits of measurements, all the calculated vertical profiles of various small components agree with the observational data. An absence of sufficiently comprehensive and reliable data on vertical profiles of concentration of minor gaseous components of the stratosphere does not permit the use of such data for testing the results of calculations obtained from modeling and selection of the most appropriate model. Insofar as the concentrations of nitric acid, the hydroxyl group and water vapor vary significantly for various models, the information on these components as most typical indicators of the adequacy of models is of maximum interest. In this case, the concentration should be determined with an accuracy of not less than 100%. It is important that the variability, introduced by an inaccurate knowledge about the rates of the Chapman cycle (O_x) and NO_x reactions, be comparatively small (not exceeding 100%). The uncertainty regarding the reaction rates of the HO_x cycle, as well as processes of interaction between HO_x and NO_x cycle introduce a maximum variability.

The problem of the rates of photochemical reactions is further complicated by their variability, caused by the effect of multiple molecular dispersion and albedo of the underlying surface. Luther and Gelinas [117], having obtained the estimate of this effect in the spectral intervals of 187–290, 290–330 and 330–735 nm, revealed the close relationship between the zenith distance of the sun, albedo at the earth's surface and its altitude above sea level and wavelength. In the first of the aforementioned spectral intervals, albedo does not affect the rate of photodissociation, since the radiation at this wavelength does not reach the earth's surface. In the second and third intervals, the combined effect of molecular dispersion and albedo can lead to changes of rate coefficients by several times at higher albedo values (i.e. above ice, snow and cloud cover). The authors of [117] have stated the necessity for the assessment of the impact of these reactions on the vertical profiles of concentration of minor components of the stratosphere. The variation of reaction rates in rela-
159 tion to the zenith distance of the sun also determines the necessity of assessing the accuracy of the assumption about the constancy of the zenith distance used in various photochemical models of the stratosphere.

Nitric oxide participating in the reactions of catalytic destruction of ozone is produced in the stratosphere as a result of reaction between $O(^1D)$ and N_2O; and nitrous oxide (just like nitrogen) is liberated in soil and water by denitrifying bacteria.

Various biological and industrial processes result in the formation of nitrogen compounds from the nitrogen present in abundant quantities in the atmosphere. Such formation of compounds from atmospheric nitrogen is known as nitrogen fixation. In nature, nitrogen compounds undergo many transformations, but in aerobic conditions characterized by the presence of oxygen, mainly compounds in the form of nitrates (NO_3^-) are formed. In anaerobic conditions, i.e. in the absence of oxygen, nitrates are denitrified and the conversion of nitrogen (contained in them) into molecular nitrogen and nitrous oxide occurs, the latter being released in the atmosphere.

Nitrous oxide, on reaching the stratosphere by way of diffusion decomposes with the formation of nitrogen along with small quantities of nitric oxide and nitrogen dioxide, which on interaction with ozone are converted into molecular oxygen. Increase in the fixation of nitrogen and the consequent increase in denitrification would lead to an intensification of releases of nitrogen oxides in the stratosphere and, consequently, a decrease in the ozone content.

The increasing amounts of nitrogen fertilizer consumption and nitrogen fixation have made it necessary to assess the impact of this factor on atmospheric ozone. In Table II.9, compiled by Pratt [136], estimates of the global level of nitrogen fixation, from data obtained during 1974, are presented. Of all the estimates, that of industrial fixation of nitrogen is most reliable. It is assumed that in 1950 the total nitrogen fixation was 174 megatons per year. The production of nitrous oxide on land is estimated at 5–10 megatons per year. Contradictory estimates for the ocean give a dispersion of 1–100 megatons per year (this upper limit is based on the assumption of supersaturation of ocean waters with nitrous oxide).

In [136] the results obtained from two methods for assessment of the impact of nitrogen fixation on the total atmospheric ozone content have been discussed. The first method is based on direct nitrogen fixation and the assumption that the relative rise in nitrous oxide production is proportional to a relative increase in total nitrogen fixation. It is also 160 assumed that sufficient time has already lapsed, so that the rate of denitrification enters into an equilibrium with the fixation (further denitrification does not lag behind the increase in fixation).

Table II.9. Approximate estimates of global level of nitrogen fixation based on data for 1974

Mechanism (source) of fixation	megatons per year	%
Natural processes on agricultural land	89	38
Natural processes in forests and on land under use	60	25
Industrial fixation in the form of fertilizers and due to other processes	57	24
Burning of fuels	20	9
Lightning	10	4
Oceans	1	—
Total	237	100

If the estimates of total fixation are based on the data of 1950, the application of the first method enables us to conclude, that an increase in nitrogen fixation by 50 and 100 megatons per year would cause a decrease in the ozone content by 5.8 and 11.5% respectively.

The second method is based on the consideration of nitrous oxide formation under the assumption that its global production reaches 100 megatons per year. This estimate is based on the continuous supersaturation of sea waters with nitrous oxide and consideration of observational data regarding the variation of nitrous oxide concentration in the atmosphere. This method gives significantly lower values for the decrease in ozone content. Thus, for example, the ozone content drops by 1% with a growth of nitrogen fixation by 100 megatons per year, if 5% of the denitrifiable nitrogen is transformed into nitrous oxide. The inadequacy of

152

the second method lies in the necessity of assuming the presence of an unknown source of nitrous oxide in the ocean and a sink in the troposphere. The production of nitrous oxide by the oceans and processes of nitrogen transformation (its global circulation) are aspects which require further investigation. The question of a shift in time between the processes of fixation and denitrification deserves serious consideration. At the same time the observational data on fluctuations of nitrous oxide concentration in the troposphere is of great importance. The potential danger to atmospheric ozone is the main reason for the urgency for such investigations.

161 Liu et al. [115] also showed that anthropogenic sources of combined nitrogen can considerably upset the nitrogen budget. Thus, for example, combustion presently produces about 4×10^7 t of nitrogen in a year. But the contribution of inorganic nitrogenous fertilizers appears to be more significant. If, in 1950 this source constituted only 0.35×10^7 t of nitrogen per year, then toward 1974, it increased to 4×10^7. Accordingly, with a 6% yearly increase the amount of nitrogen from the above source would constitute 20×10^7 t per year toward 2000. From the 1974 data, the total nitrogen fixed on land was 26×10^7 t per year, i.e. the contribution of anthropogenic sources should be considered as significant. This is a warning signal regarding the possible ecological consequences of anthropogenic production of nitrogen.

The problem of the impact of anthropogenic production of nitrous oxide on the ozone layer, and in this connection, the assessment of nitrous oxide releases into the atmosphere, its residence time in the atmosphere, the intensity and sign of oceanic sources of nitrous oxide has drawn special attention. If we take into account the total nitrous oxide content in the atmosphere and assume that nitrogen production from oceans constitutes 8.5×10^7 t per year, then the residence time of nitrous oxide in the atmosphere should be 20 hours. Insofar as the data on the sink of nitrous oxide in the stratosphere leads to the conclusion that its residence time is about 150 years, this indicates the presence of some unknown sink of nitrous oxide in the troposphere or on the earth's surface.

McElroy et al. [122, 159], indicated, that over the last 25 years, industrial fixation of nitrogen for agricultural application has increased by more than 10 times (up to 3.5×10^6 t per year), and comprised more than 2% of the global natural fixation of nitrogen according to data for 1950 (1.6×10^8 t per year). The annual growth of industrial fixation during 1950–1974 was 10.7%. Extrapolation to the end of the century leads to values of industrial fixation of nitrogen at $(1–2) \times 10^8$ t per year. This confirms a noticeable anthropogenic effect on the global nitrogen cycle at present and a rapid intensification of this effect in the near future may be anticipated.

One of the manifestations of the anthropogenic effect on the nitrogen cycle is the increase in release of nitrous oxide into the atmosphere, which (with the consideration of the aforementioned data on industrial fixation of nitrogen) may go up in the near future by approximately three times. The main source of nitrous oxide is denitrification in the soil. Although reliable information on the contribution of the ocean to the global nitrous oxide budget is not available, it may be considered that its continental resources (soil, groundwater and estuaries) are most significant. All sources of nitrogen in a combined form, either mineralized in soils or carried in the form of fertilizers, ultimately release it into the atmosphere in a molecular form.

The temporal scales of denitrification closely depend on the peculiarities of agricultural practices, regional and climatic factors. However, rapid and slow denitrification, determined by characteristic periods of the order of 1–10 or 10–100 years, should be distinguished. Denitrification should occur relatively faster, particularly in those countries where the major agricultural crop is rice, which accelerates the anaerobic processes. It is assumed that on the average, about 50% of the combined nitrogen, introduced into the soil in the form of fertilizers, denitrifies over a period of 10 years with an average lag of 7 years with respect to the release of nitrous oxide in the atmosphere. About 40% of the combined nitrogen is freed after 50 years, and an equilibrium on the whole is established within approximately 100 years.

The calculation of global dynamics, carried out in [122], of nitrogen oxide deliveries and the impact on the ozone layer showed that during the second half of the twenty-first century, the drop in the total ozone content, as a result of intense use of mineral fertilizers, would be 20%. In this connection, the authors of [122] have criticized the work of Crutzen [65] in which a decrease in ozone content toward the end of the twenty-first century by less than 10% was predicted. However, Ellsaesser [76, 78, 80] has advanced certain arguments in favor of the view that the anthropogenic impact on the global nitrogen cycle has little significance.

The following factors also support the formation of nitrous oxide: 1) decrease in oxygen concentration; 2) increase in humidity (this is responsible for the elimination of oxygen; 3) increase in pH; and 4) decrease in the concentrations of nitrites and nitrates. In the process of land cultivation, because of various factors there had been a gradual increase in the oxygen content of the soil, a decrease in soil moisture which promotes an increase in oxygen level, and an increase in the pH. Simultaneously, there were effects of the opposite sign, promoting the formation of nitrous oxide as a result of irrigation, rice cultivation and increase of nitrogen fixation in the soil. But these effects usually appear on much smaller spatial scales.

154

Often the assumption is made that gaseous sulfur and nitrogen of anthropogenic origin cause a large-scale increase in the acidity of precipitates and intensify nitrous-oxide formation. These theoretical conclusions are not confirmed by observational data. On the contrary, natural releases of gaseous sulfur and nitrogen considerably exceed those of anthropogenic origin.

It is probably true that any anthropogenic impact on denitrification, which has already occurred or may arise in the future, tends to decrease

163 the formation of nitrous oxide and, consequently, promotes an increase in the ozone content of the atmosphere. The observational data on the trends of nitrous oxide concentration are contradictory, but the results of measurements of the total ozone content indicate a global trend toward increase from the moment of the first measurements (in the early twenties) and especially after a considerable expansion of the observational network in the mid fifties.

The aforementioned assessment, has been obtained without taking into account the gigantic global pool of combined nitrogen on land and in the oceans. This means that the time lag between the introduction of fertilizer in the soil and a significant manifestation of denitrification may be several hundred or even a thousand years. On the basis of the obtained data on the production of nitrogen fertilizers, Liu et al. [115] arrived at the conclusion that the increase in the proportion of nitrous oxide in the mixture would reach 4% by the year 2025 and 20% by 2050. This may cause decreases in the ozone content by 1% and 4% respectively. If we assume, that there would always be a balance between nitrogen capture and denitrification, the increase in the concentration of nitrous oxide would be 220% by the year 2025 and approximately 10 times by 2050. If the production of fertilizers is maintained at the level of production in 2025, then after a lapse of 1000 years the concentration of nitrous oxide would increase by 40 times, i.e. for such time scales the consequences of fertilizer application may be very serious.

Insofar as the rapid manifestion of the impact of nitrogen fertilizer application is concerned, this is possible only with the existence of a mechanism of denitrification, which is a hundred or even a thousand times more effective than natural biological processes. Such a mechanism is still unknown, and in this connection the study of reactions of the denitrification process on the rapid increase in the formation of combined nitrogen in the soil over limited territories is of great interest.

3.3. Anthropogenic impact on ozone layer and climate. Table II.10 has been compiled from the information appearing in the review article [97]. The table characterizes the extent of the impact of supersonic aviation and other anthropogenic factors on ozone and climate. This information

155

Table II.10. Possible anthropogenic effect on ozone and climate (question mark denotes uncertain conclusions)

Pollutants and their effect	Source of pollution				
	aviation	refrigerants, aerosol dust and other sources of Freon	Power stations (fossil fuel)	nitrogen fertilizers	atomic power stations
1	2	3	4	5	6
Pollutants	NO_x[1], H_2O[1], SO_2, soot, particles, CO_2	Chlorofluorocarbon compounds	CO_2[1], aerosol N_2O, NO_x, H_2O heat, SO_2	N_2O[1], NH_3 (?)	Radioisotopes, (^{85}Kr)[1], ^{3}H, ^{14}C, heat
Effect on ozone	Increase or decrease	Decrease (potentially large)	Small increase, if contribution of N_2O is small	Decrease (?)	Decrease (?) (formation of NO)
Effect on climate	Warming (?)[2]	Warming[3]	Warming (potentially high)	Cooling (?)	?
Level (altitude of pollution)	6–20 km	Surface	Surface	Surface	Surface
Altitude of pollution sink	About 30 km	30–40 km (initially); about 30 km (finally)	Surface (oceans, vegetation cover)	Stratosphere (initially), about 30 km (finally)	Disintegration, distributed over space
Reaction time (restoration of properties) of atmosphere, years	1–5	50–100	100+(?)	10–100 (+)	Decades

(Contd.)

156

1	2	3	4	5	6
Specific aspects and uncertainties	Altitude of tropopause, transfer, chemistry of upper troposphere, concentration of sources wrt latitudes, appearance in two or three dimensions, forecast on emissions	Tropospheric sinks (if existing)	CO_2 budget, equilibrium with oceans, geographical after effects	Unknown sources of N_2O, time of establishment of equilibrium for N pools	Atmospheric electricity, impact of ions on clouds, transfer
General aspects	Chemistry of atmosphere, transfer, climatic phenomenon effect on biosphere (with variation of ozone)	Chemistry of atmosphere, transfer, effect on biosphere and climate	Climatic phenomenon	Chemistry, transfer effect on biosphere	Not clear

[1] Most important component.
[2] Multicomponent impact: cooling because of aerosol and decrease of ozone; warming caused by water vapor; by inversion traces and nitrogen dioxide.
[3] Contribution of 'greenhouse' effect predominates.

can serve as a guideline in solving the problems of anthropogenic impact on the ozone layer and climate.

As already mentioned, the main difficulty in numerical modeling of the impact of anthropogenic and natural origin on stratospheric ozone lies in the necessity to account for various kinds of interactions, as well as the influence of the oxygen, hydrogen, nitrogen and chlorine cycles, which determine the ozone budget. Table II.11, compiled by Bauer [40], characterizes the various factors having impact on ozone which should be taken into account, and the variability of these factors with time during 1950-1960. The eruptions of the Bezymyannyi, Puiekh, Agung, Sheveluch, Avu, Fernandina, and Fuego volcanoes have been ranked as massive. In Table II.12, certain approximate data on the quantities of ejected materials has been presented.

For comparison it should be noted that the 'background' content of $NO_y = NO_x + HNO_3$ ($NO_x = NO + NO_2$) constitutes $(4-15) \times 10^{34}$ molecules; $HO_x = OH + HO_2$ about $(15-30) \times 10^{32}$ molecules, and $ClX = Cl + ClO + HCl + ClONO_2$ about 10^{34} molecules. While estimating the injection of NO_x into the stratosphere during nuclear tests, it is assumed that they constitute 1×10^{32} molecules for each megaton explosion. Among halocarbons, $CFCl_3$ (Freon-11), CF_2Cl_2 (Freon-12) and CHF_2Cl (Freon-22), as well as 10 other halocarbon compounds (carbon tetrachloride, chloroform, methyl chloroform, perchloroethylene, trichloroethylene, etc.) have been taken into account.

The estimates of effusions due to volcanic eruptions are highly approximate. Of these only the most prominent have been taken into account: Bezymyannyi (March 30, 1956); Puiekh (May 21-24, 1959); Agung (March 17, May 5, 1963); Sheveluch (November 12, 1964); Avu (August 12, 1966); Fernandina (June 11-12, 1968); and Fuego (October 10-23, 1974). As far as the galactic cosmic rays and solar proton events are concerned, it is assumed that each pair of ions formed under the impact of particles in the energy range of 30-500 Mev produces 1.3 molecules of nitric oxide and 2.0 molecules of the hydroxyl radical. In this context it is known that the maximum of the galactic cosmic rays coincides with the minimum of the 11-year sunspot cycle, and the major solar proton events occur 1-5 times during the same cycle and continue for 1-3 days. The earth's magnetic field limits the zone of appearance of galactic cosmic rays and solar proton events to higher latitudes.

If we compare the background contents of NO_y and HO_x with those formed because of galactic cosmic rays and solar proton events, it becomes clear that the effect of ionization caused by cosmic rays is considerably less significant as compared to other factors (releases of carbon halides, volcanic eruptions, nuclear tests). In the case of anthropogenically caused fixation of nitrogen ($\gtrsim 75\%$ fixation is linked with the

158

Table II.11. Possible factors of natural and anthropogenic effects on the ozone layer (plus sign indicates the presence of corresponding phenomenon)

	Year													
	1950	1952	1954	1956	1958	1960	1962	1964	1966	1968	1970	1972	1974	1976
Natural geophysical phenomenon														
11-year cycle of sunspots			Min		Max			Min		Max				Min
Galactic cosmic rays			Max		Min			Max		Min				Max
Prominent solar proton events						May-July, Nov								
Maximums of quasi-biennial oscillations				+					+	+	+	+	+	+
Spontaneous heating of stratosphere				+						+	+			
Prominent volcanic eruptions				+					+					
Anthropogenic factors														
Nuclear explosions in atmosphere (power, megatons per year)		60		30	90	120	220							
Carbon halides		Releases of chlorofluorocarbons with annual increase of 10% (1955–1975)												
Industrial fixation of nitrogen		Annual increase by 10% (1955–1975)												
Burning of fossil fuel		Annual increase by 5% (1955–1975)												

Table II.12. Possible deliveries of various components

Source	Example of effect on ozone	Release (molecules)	Altitude of delivery, km	Region
Galactic cosmic rays	Difference in contents of components of solar maximum and minimum	7×10^{32} HO_x/year 5×10^{32} NO_x/year	10–30	Polar regions (latitudes above 60°)
Solar proton events	5–7 August 1972	4.5×10^{33} HO_x 3×10^{33} NO_x	30–35	—do—
Prominent explosive volcanic eruptions	Mount Agung 1963	9×10^{33} ClX, H_2O (about 2% of usual level). Dust; most severe increase as compared to background. 2×10^{34} atoms of S in the composition of SO_2 or H_2SO_4	16–30/35	8.5°S
Nuclear testing in atmosphere	Autumn 1962 (180 megatons)	$(0.9–1.8) \times 10^{34}$ NO_x	10–35	77°N
Carbon halides	1973	7×10^{33} ClX/year	Earth's surface	Mainly 30–55°N
Industrial fixation of nitrogen	1973	2×10^{35} combined atoms of nitrogen in a year	Earth's surface	—do—

production of agricultural fertilizers), it is important to take into account the considerable delay in the process of denitrification of combined nitrogen (about 6% of the total nitrogen is transformed into N_2O, the remainder into N_2) and subsequent transformation of nitrous into nitric oxide, which enters into reaction with ozone. The total delay, apparently, exceeds 10 years.

169 In spite of the undoubtedly significant role of all the aforementioned effects related with variations of chemical composition of the stratosphere, it is entirely probable that the variability of extraterrestrial ultraviolet solar radiation in the course of the 11-year cycle plays a major role. Such effects are manifested in several ways: 1) the processes of photodissociation, which primarily lead to the formation of ozone, 2) localized heating due to absorption of ultraviolet radiation, and 3) changes in temperature, dynamics and photochemical processes as a result of this heating. The data of satellite observations demonstrates that the amplitude in the 200–300 nm wave band from the maximum to minimum of the 11-year cycle attains 20–50%, whereas the variability in the region of shorter wavelengths (170–200 nm and the L_α line) is even more significant. If these results are correct (they need to be confirmed), the overall ozone content should increase approximately by 8% during the period from the maximum to minimum of the 11-year cycle, which considerably exceeds the contribution of all other factors contributing to the impact on ozone during the peroid 1955–1975. Therefore, prolonged (at least during two 11-year cycles) and precision monitoring of extraterrestrial ultraviolet solar radiation is very important.

From the viewpoint of the data on the trend of water-vapor content in the stratosphere which plays an important role in the processes contributing to impact on ozone, the situation remains very unsatisfactory. Hence, continued observations on water-vapor content in the global stratosphere are highly important. A recent assessment of the contribution of various factors toward the effect on ozone leads to the conclusion regarding the insignificance of releases from supersonic aircraft and a considerably weaker effect of nuclear tests than was earlier assumed. However, the evolution of estimates characterizing the role of carbon halides appeared to be the opposite (according to latest estimates the effect is twice as great as in 1976). This attracts special attention to the problems of carbon halides, and also toward volcanic eruptions as possible sources of release.

Insofar as the concentration of ozone in the stratosphere undergoes a geographical seasonal and year-to-year variability, which may be subjected to the influence of different trace elements in the stratosphere, it is important to study the effect of this variability on climate. The complexity of the ozone-climate problem is determined by the possibility of a direct (due to radiation), as well as an indirect (dynamic) effect on the climate of the troposphere.

The radiative interaction between the troposphere and stratosphere manifests itself as a result of the effect of ozone variability on radiative transfer. This relates to the absorption of solar radiation by ozone in the visible and ultraviolet regions of the spectrum, which determines the mo-
170 dulation of a flux of solar radiation reaching the troposphere, as well as radiant heat exchange: the stratosphere is warmed because of absorption of the outgoing flux of long-wave radiation in the 9–10 μm wave band and emission from ozone in the same interval contributes to the energetics of the troposphere. Since ozone absorbs more radiation than it emits, a decrease in ozone content should lead to a cooling of the stratosphere and a decrease in the energy being transferred to the troposphere because of ozone release.

Dynamic interaction between the troposphere and the stratosphere caused by ozone, is manifested particularly through a change in the conditions of propagation of global waves from the troposphere into the stratosphere. This occurs under the influence of variations in the temperature field of the stratosphere caused by a variation in the ozone content. A sequential analysis of the impact of ozone content variability on climate requires numerical modeling on the basis of a three-dimensional model of climate. However, the complexity of such a formulation of the problem determines the merits of the study, in the first place, of that part of the problem which is related with the effect of ozone on the radiation budget of the earth-troposphere system.

Ramanathan and Dickinson [141] carried out calculations for the 24-level model of the atmosphere (0–54 km) for four seasons in the northern hemisphere and for the following three hypothetical cases of variability of ozone concentration in the stratosphere: 1) a uniform decrease of 30% at all latitudes; 2) an altitudewise redistribution of concentration while preserving the total content; and 3) disturbance of the concentration field under the effect of chlorofluoromethane calculated on the basis of the photochemical model with the assumption that chlorofluoromethane will be maintained at the contemporary level over an unlimited period.

In Figure II.8, mean annual values of the radiation budget for the unperturbed stratosphere and its components for the northern hemisphere, illustrating the existence of a radiative equilibrium, are presented. An analysis of data given in this figure showed that the relative contributions of O_3 and $CO_2 + H_2O$ in the total radiation budget appear to be quantitatively almost the same, though opposite in sign. It follows from the figure that shortwave radiation absorbed by the stratosphere is almost half that of absorbed long-wave radiation. Analysis of the annual trend

162

shows that absorption of long-wave radiation by carbon dioxide and water vapor promotes a warming of the lower stratosphere in polar latitudes during summer.

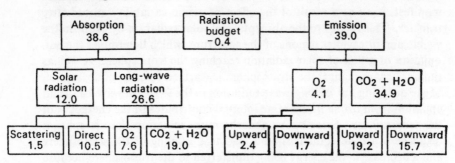

Fig. II.8. Mean annual radiation budget of the stratosphere for the northern hemisphere (W/m^2).

The impact of the stratosphere on the radiation budget of the troposphere has been assessed with the consideration of the fact that the albedo of the troposphere is equal to 0.31 and half of the long-wave radiation in 171 the carbon dioxide and water-vapor bands is formed because of energy liberated in the absorption of solar radiation by ozone. The total contribution of the stratosphere in the radiation budget of the troposphere on account of shortwave and long-wave radiation is positive and constitutes $10.7 \ W/m^2$ with the share of ozone amounting to $2.3 \ W/m^2$. The latitudinal distribution of the radiation budget of the stratosphere is characterized by heating in the latitudinal zone from 0 to 35° and cooling at higher latitudes, which give rise to a temperature inversion under the influence of dynamic factors. The impact of radiation on the troposphere, caused by a decrease in the ozone content of the stratosphere, is manifested in two ways:

1) as a result of the effect of solar radiation flux and (less in quantity) decrease in long-wave radiation of the stratosphere; and

2) as a result of decrease in long-wave radiation because of a drop in temperature of the stratosphere.

In the case of a uniform decrease in ozone concentration, a very small decrease in the tropospheric radiation budget takes place, which on the average over a year varies from less than $0.1 \ W/m^2$ at lower latitudes to more than $0.4 \ W/m^2$ at higher latitudes. In this the direct radiant heating is concentrated at lower latitudes and close to the surface, while cooling caused by a drop in temperature of the stratosphere, is comparatively more uniformly distributed in the layer of the upper troposphere.

Analysis of the results obtained by varying the form of vertical profiles of ozone concentration (for its unchanged total content) and latitudinal

distribution (for a zero mean variation) enabled us to conclude, that even a slight variation in the profile (about 5% of the total ozone content) causes a variation in ozone content comparable to a 30% uniform
172 decrease in the ozone concentration when there is a latitudinal variation. Any increase (decrease) in ozone content in the lower stratosphere accompanied by a compensatory decrease (increase) at higher latitudes can cause variations in the radiation budget of the troposphere. These changes exceed, in order of magnitude, the values of observed changes during a uniform decrease in concentration. The above conclusions demonstrate the importance of consideration of redistribution of ozone along altitude, as also its latitudinal variability and annual trend in the assessment of the climatic impact of ozone.

An important result of calculations is as follows: changes in the radiation budget of the troposphere, caused by variation in the ozone content in the stratosphere, appear to be different not only in magnitudes but also in sign when cooling of the stratosphere accompanying the decrease in ozone content is taken into account (in this case the radiation budget of the troposphere decreases by 0.16 W/m^2 instead of increasing by 0.59 W/m^2). Estimates of the impact of chlorofluoromethane approximately corresponding to a 19% uniform decrease in the ozone concentration (here a decrease in ozone concentration in the troposphere, causing a 6% drop of the total ozone content and a lowering of ozone concentration to 25–50% in the 39–50 km layer is an exception) showed that a variation in the radiation budget of the troposphere is equivalent to changes taking place with a uniform decrease in ozone concentration. The effects of 'anomalous' variability in the troposphere and upper stratosphere are mutually compensatory.

Though the troposphere contains about 6–14% of the total ozone content, the variability of ozone concentration in the troposphere should be taken into consideration as a factor contributing to the greenhouse effect of the atmosphere. From the standpoint of the climatic impact of depletion in the ozone content, it is important that a small variation in the radiation budget of the earth-atmosphere system should produce a considerable warming of the layers in the vicinity of the earth's surface with a cooling in the upper troposphere.

Dickinson [69], while discussing the most important areas of development in atmospheric sciences during the next 20 years, indicated that these are linked with the investigations in the field of atmospheric chemical reactions and their impact on climate and the biosphere, as also with the study of long-term changes in climate. The two problems mentioned here are chiefly solved by means of observations from satellites in combination with a theoretical treatment of the data. In achieving this, on-line processing and analysis of observational data are of exceptional importance.

164

The maximum attention in the field of stratospheric chemistry is given to the problem of ozone. As already mentioned, the investigations carried out during recent years enabled us to draw conclusions about the 173 minor importance of nitrogen oxides, along with the more significant role of the chlorine cycle, on changes in ozone. The principal natural component in the nitrogen cycle is nitrous oxide formed during the process of denitrification in soil and oceans by bacteria. The anthropogenic factors, contributing to the production of nitrogen oxides, are fertilizers and supersonic aviation. The contribution of the chlorine cycle of ozone destruction is determined by the chlorofluorocarbons released due to anthropogenic factors.

The most important problem of the near future is to adequately understand and to quantitatively assess the processes governing the ozone-layer formation in all its complexity, and also the consequences of variation in ozone content *vis-a-vis* climate and the biosphere. Theoretical models, taking into account the influence of all the factors on ozone, would presumably be developed toward the end of the eighties, and global observations on the most important components of the stratosphere affecting the ozone-layer formation would be recorded. Investigations concerning the consequences of ozone-content variation on climate and the biosphere would, obviously, require more time. In the problem of climate the study of anthropogenic effects is most important and any investigation on various physical processes affecting the climatic changes would require the utmost attention.

4. THE NATURAL ENVIRONMENT AND NUCLEAR WAR

Among the various factors that necessarily influence the composition of the stratosphere, one cannot fail to mention nuclear tests. In view of this during recent years, many publications devoted to the assessment of the possible influence of a nuclear war on the environment, have appeared. The US National Academy, published the report *Long-term Effects of Multiple Nuclear Weapon Detonations* compiled by a group consisting of eight specialists (with the participation of 48 other experts) under the chairmanship of Neher (from the University of Minnesota). The principal conclusion of the report is that the impact of a total nuclear attack on all countries, apart from those subjected to direct attack, will not be as catastrophic as many scientists have predicted [43]. In view of the seriousness of such a conclusion, Handler, the President of the Academy, enclosed a letter along with the report in which he indicated that factors beyond the purview of investigation of the Academy lie behind the 'total unpredictability' of consequences of a nuclear holocaust.

In view of the contradiction inherent in the aforementioned report and the letter, the Federation of American Scientists (FAS) came forward with a criticism of the report by the US National Academy. The FAS 174 implicitly accused the US National Academy of its premeditated approval of nuclear warfare and, in fact, considered its conclusion regarding the minimal consequences of a widespread application of nuclear weapons an erroneous judgment.

The US Arms Control and Disarmament Agency, perturbed by the tone of the report prepared by the Academy on contract with this agency, has also issued a statement. The Agency asserts that the information contained in the report, in fact, demonstrates that no aggressor can start a nuclear war, since it would lead to disastrous economic and ecologic consequences, even for the aggressor.

Two days after the publication of the report and letter, Handler expressed his concern in a journal, 'Science' in which he clarified that in his view a nuclear holocaust would be so destructive that no one would be able to escape its disastrous effects.

In the said report of the US National Academy, the possible consequences of nuclear explosions, carried out in the northern hemisphere, with a total capacity of up to 10000 megatons (this corresponds to half of the stockpiled nuclear weaponry) for those countries which were not subjected to direct attack are discussed. In this, chief attention is paid to an analysis of such consequences as are retained over a period of approximately 30 years. The conclusion arrived at is as follows: though a global nuclear attack would cause extensive destruction, its consequences would not be catastrophic on a global scale over a decade, because mankind would be in position to survive a nuclear holocaust.

Contradicting these views the FAS stated that contemporary science is not in a position to assess with sufficient accuracy the biological and ecologic consequences of explosions of 10000 megatons. It should be emphasized that this view of the FAS is more realistic.

From the specific conclusions contained in the report of the Academy, the following appear to be the most important:

1) release of nitrogen oxides into the stratosphere as a result of nuclear explosions can lead to a decrease in the total ozone content by 30–70% with subsequent restoration of its normal content after 2–4 years;

2) release of nitrogen oxides and dust may result in a decrease in temperature amounting to several degrees, or a very slight warming (most probably, the climatic aftereffects would not exceed the usual limits of variability);

3) in spite of the initial significant impact of explosions on the ecosystems, the latter would be sufficiently well-restored after 25 years;

4) the impact on agriculture could be significant (especially due to intense ultraviolet radiation); and

166

5) reinforcement of ionizing radiation could cause damage to the biosphere, a rise in the incidence of cancer, and genetic diseases.

In [36, 38, 52, 56, 70, 87, 99, 104, 129, 174], the possible consequences of nuclear explosions from the standpoint of influences on the atmosphere, which may appear in global dimensions over a period of several decades, have been discussed. It is assumed that up to half of the accumulated nuclear weapons would be used (500–1000 explosions each of which is equivalent to 10–20 megatons of TNT, as well as 4–5 thousand explosions of 1–2 megatons, i.e. the combined capacity of the explosions is equivalent to 10^4 megaton). In these estimates, it was assumed that the points of explosions would be uniformly distributed in the 30–60°N belt. In this case two aspects of the problem have been considered:

1) the influence of the atmosphere in the propagation of the products of nuclear explosions; and

2) the impact of explosions on the atmosphere and the related consequences (particularly, the possible influences on climate).

Earlier works showed that the zonal transport of products of explosion to the stratosphere is very fast and attains global dimensions over a period of a few weeks. Significantly slow is the meridional transfer (several months for covering the entire hemisphere). Vertical transport is still slower: from several weeks to several months in the vicinity of the tropopause and up to several years in the middle and upper stratosphere.

An analysis of diffusion processes of radioactive products of explosions conducted in the northern hemisphere led to the conclusion, that due to the transport of short-lived components to the troposphere their fallout in individual sections should occur in high concentrations of radioactive substances. Insofar as leaching by precipitates makes the principal contribution in the cleaning of the troposphere, the maximum concentration of radioactive substances should occur in the 'green belt' with its intense precipitation.

The area covered by the fallout of explosion products is considerably more extensive in the stratosphere. In this case the radioactivity of short-lived components may, by the time they fall to the earth's surface, decrease to a very low level. The concentration of radioactive products in the lower stratosphere (at an altitude of 20 km) should drop by a factor of e, roughly over a year. Less than half (probably, about 1/3) of the stratospheric radioactive products should penetrate into the stratosphere of the southern hemisphere. Therefore, the maximum fallout of the products of explosions over the earth's surface in the southern hemisphere would constitute about one-third of the amount in the northern
hemisphere. The mean surface radioactivity in the northern hemisphere would be about 1 c/km².

As a result of explosions of 10^4 megatons, each one of which is of more than one megaton, a large amount of nitrogen oxide would move into the stratosphere. In terms of nitric oxide it would be 10^{36} molecules, which exceeds by 5–50 times the background content of nitrogen oxides and should cause a considerable decrease in the total ozone content of the northern hemisphere. The approximate estimates, based on the use of a one-dimensional model of the atmosphere, confirm that just after the explosions, the ozone content may drop by 30–70%. The existing values for the stratosphere of the southern hemisphere are significantly lower, and therefore, the mean global decrease would be 20–40%. It should be assumed that over 2–4 years the loss of ozone would be recovered to the extent of 60% through natural processes, and after 10 years, the lowering in the ozone content caused by explosions would be 1–2%.

These estimates are however, purely approximate: all the indicated values have been found with an accuracy up to a multiple of 2–3. More reliable estimates can be obtained on the basis of three-dimensional modeling, which is essential for taking into account the changes in the dynamics of the stratosphere caused by depletion of the ozone content. Data on the variation of ozone content after the nuclear tests can serve as a criterion of the accuracy of the said estimates. Some attempts to discover such consequences of thermonuclear tests in 1961–1962 (assuming that the release of nitric oxide in this case constituted $(0.4–1.5) \times 10^{32}$ mol/megaton, and the maximum decrease of ozone content could reach 4–5%) did not indicate positive results [56], but neither can they be said to definitely reject similar consequences (the solution of the problem becomes complicated, particularly because of the difficulty of reliably eliminating the effect of the quasibiennial periodicity). A reliable recording of the reduction in ozone content could not be achieved from data of the *Nimbus-4* satellite during the periods of nuclear tests of 1970–1974.

Calculations carried out by Folly and Ruderman in 1973 showed that nuclear tests in the atmosphere toward the end of the fifties and beginning of the sixties should have caused depletion of the total ozone content by more than 10%. On the other hand, estimates of Chang and Duewer, obtained on the basis of a one-dimensional nonsteady model led to the conclusion that the mean annual decrease of ozone content in 1963 should have constituted about 4%. According to data obtained by Johnston the drop in ozone content should not be more than 5%.

177   The attempt to discover the variation in ozone content, caused by nuclear tests, on the basis of observational data did not provide positive results due to the impossibility of distinguishing the predicted changes against the background of natural variations. Thus, for example, Johnston while analyzing the global observational data for 1960–1970 discovered a statistically insignificant decrease of the total ozone content by 2.2%

during 1960–1962 followed by a statistically significant increase of 3.7% during 1963–1970. This increase was explained as restoration of the ozone content after a decrease of the order of a few percent caused by nuclear tests. Other data indicates that the said decrease preceded the tests, and the increased ozone content appears to be higher than the average over many years. Angell and Korshover suppose that the effect of nuclear tests on ozone does not exceed 1–2%.

An analysis of the impact of nuclear tests on ozone requires consideration of the contribution of such factors, as solar activity, volcanic eruptions and chlorofluorocarbons, not yet accurately estimated. This indeterminacy impelled Chang, et al. [52] to obtain new theoretical estimates of the impact of nuclear tests on ozone within the framework of a one-dimensional model with the use of a considerably modified model of photochemical processes in the stratosphere, which takes into account about 100 reactions.

From an analysis of the data on the deliveries of nitrogen oxides (NO_x) into the stratosphere during explosions, the extent of release has been assumed at 0.67×10^{32} mol/megaton of TNT with an accuracy of about $\pm 50\%$. It is assumed that the concentration of explosion products increases proportionally with altitude within the limits of the cloud formed after explosion. The lower boundary of the cloud lies at temperate latitudes at a level (in kilometers) of $13.41\ Y^{0.2}$, where Y is the power of the explosion (megatons) and the upper boundary at the level of $21.64\ Y$. At higher latitudes, a correction is applied, increasing the lower and upper limits of the boundary by 4 km, and in the tropical belt, decreasing it by 2 km.

It has been assumed that the products of explosion are uniformly distributed over the stratosphere of the northern hemisphere. The photochemical model takes into account the HO_x, NO_x, O_x, ClO_x cycles, and the most reliable data on reaction rates (calculations showed that effects under assessment are very sensitive to the selection of the photochemical model). While considering the chlorine cycle, it is assumed that the concentration of ClO_x is 1.1–1.3 ppb, and chlorine and oxides containing chlorine are formed from methane and hydrogen tetrachloride having been assigned a constant concentration in the minor gaseous components under consideration.

Analysis of the results obtained from numerical modeling showed that a reduction in the total ozone content in 1963–1964 had a weak relationship with vertical profiles of the coefficient of mixing. By using a new (1977) photochemical model of restoration of the ozone layer after nuclear tests, the rate of vertical redistribution of products of delivery is first determined, and only at later stages of numerical modeling is it controlled by the rate of removal of excess nitrogen oxide from the stratosphere.

The combined effect of a number of modifications in the photochemical model of 1973 was an improvement in the maximum reduction of the ozone content in January 1963 to 8.2% (as compared with 4.9%, according to results of calculations for the 1973 model). With an additional consideration of the chlorine cycle and modification of the hydrogen cycle, the estimates of impact of nuclear tests were found to be 30% lower, but all the same significantly higher than those obtained by Angell and Korshover from observational data.

Computations relating to 1963 (with consideration of all the tests conducted till that time) show that the mean annual drop of ozone content in the northern hemisphere is 3%. Hence, the new one-dimensional model enables us to obtain estimates of the impact of nuclear tests on ozone that are in agreement with observational data.

From the standpoint of the obtained results investigations made during the sixties can be a useful source of information for the testing of theoretical models of the stratosphere in the negative sense only, as they enable us to reject any model which leads to a depletion in the ozone content, under the influence of tests, considerably exceeding 2%. The modification in the model of photochemical processes without affecting the nitrogen cycle led to very significant changes in the earlier obtained estimates. Hence, further modification of theoretical models of the stratosphere within the framework of two- and three-dimensional modeling, as well as continuation of experimental investigations on the rates of photochemical reactions are essential.

As already indicated, the use of the latest measurement data on the rates of photochemical reactions for assessment of the stratospheric ozone budget enabled us to draw the conclusion of a considerably low sensitivity of ozone to releases of nitrogen oxides NO_x ($NO + NO_2$) in quantities comparable with the natural concentration of NO_x. The lowering of sensitivity is due to the interaction between the NO_x and HO_x cycles of catalytic destruction of ozone. In the case of intense injection of nitrogen oxides into the stratosphere, similar to those which may arise in the case of a nuclear war with widespread employment of weapons with capacities of more than one megaton, the intensity of injection may appear to be more than two orders higher than in the case of anthropogenic effects discussed earlier (supersonic aviation, chlorofluorocarbons, fertilizers). In this case the saturation processes causing a reduction in the ozone content may take place, all the same, the depletion of the ozone layer should be heavy. On the other hand, the influence of submegaton explosive charges or releases at lower altitudes should be less significant. Duewer et al. [70] obtained the estimates of the effect of massive sporadic discharges of nitrogen oxides in the 9.5–16.5 km and 17.5–28.5 km layers on ozone by using a one-dimensional photochemical model of the stratosphere. The first layer corresponds to the level of stabilization for cloud formed by 0.25 megaton explosion near the earth's surface and the second layer signifies the assumed level of stabilization in the case of a 4 megaton explosion.

According to available estimates, the total release of nitrogen (in the calculation for 1 megaton) is equivalent to 6.7×10^{31} molecules of nitrogen oxide and thus may be considered to be uniformly distributed within the range of the northern hemisphere. Numerical modeling of the impact on ozone has been carried out under the assumption that the total capacity of explosions for each of the aforementioned layers is equal to 100, 1000, 5000 and 10000 megatons.

In the case of extremely copious releases into the stratosphere (10^4 megatons with the use of 1–4 megaton charges), a very severe depletion of the total ozone content takes place immediately after the explosion. This reduction may be as much as 30–60%. Restoration of the ozone content takes place after approximately 4–5 years. However, even the most powerful explosions at altitudes of less than 15 km cause only a small (of the order of a few percent) variation in ozone content (mostly positive, which is related with the mechanism of ozone formation, similar to the mechanism of 'smog' formation).

A very weak effect of explosions on ozone in the lower stratosphere and troposphere is due to considerable leaching of nitric acid and nitrogen dioxide from the troposphere, as a result of which the amount of nitrogen oxides reaching the middle layers of the troposphere appears to be comparatively small. In such conditions the smog effect of ozone formation in the troposphere and lower stratosphere, is dominant.

In the case of a heavy impact of explosions on the stratosphere, the nitrogen oxides, propagated after explosion in the upper and lower layers of the atmosphere, cause a decrease in the ozone concentration at higher altitudes, and its increase in the lower stratosphere and troposphere. For more reliable future estimates, it is essential to take into account the interaction between changes in chemical composition in the stratosphere and the dynamics of the atmosphere. In the case of severe effects, being studied, such interaction may appear to be significant. No doubt, the determination of reliable values of the chemical reaction rates plays an exceptionally important role.

One of the serious consequences of ozone depletion in the atmosphere is the rise in the intensity of ultraviolet radiation. If we take into account a depletion in the ozone content by 50% in the northern hemisphere as typical for the first year after a nuclear explosion, the intensity of ultraviolet radiation at wavelengths of 292.5 and 302.5 nm increases at midday by 25 and 2.5 times (20°N) and by 120 and 4 times (50°N) respectively.

The variations in the composition of the stratosphere related with depletion of ozone, increase in the amount of nitrogen oxides and dust content of the stratosphere can lead to a variation of climate, caused by corresponding changes of radiant heat influx. Rough estimates of the global climatic impact can be obtained from the one-dimensional model of radiative-convective equilibrium. As a result of explosions of a total of 10^4 megaton capacity, about 10^7–10^8 t of dust is injected into the global stratosphere, which is approximately equivalent to dust loading of the stratosphere after the eruption of Mount Krakatau in 1883 and should reduce the influx of solar radiation at the level of the earth's surface by a few percent over a period of 1–3 years. This, in its turn, should cause a lowering of the mean global temperature of the underlying surface by a fraction of a degree.

In the case of nuclear explosions, the possibility of temperature variations, which may have consequences other than a mere change in sign, is not excluded. Major changes in ozone content should give rise to considerable variations of the stratospheric temperature, which apparently, will extend into the troposphere. If it is assumed that the mean global depletion of ozone is 50% and is uniform at all altitudes then approximate estimates show that the associated drop in the mean global temperature of the underlying surface would be a fraction of a degree. Due to the approximate nature of calculations, their modification can lead to changes in their values, and also in the sign of temperature variation.

A possible factor affecting climate is the variation of albedo of the underlying surface caused by reallocation of agricultural zones after 'nuclear exchange'. It is important to mention that comparatively small changes in mean global temperature may be accompanied by significantly greater variations of regional climate in temperate and higher latitudes, having an importance from the viewpoint of agriculture (for example, decrease in the number of days with no frost, etc.). Since similar variations are within the limits of the natural variability of climate, the time required for their damping should not exceed a few years. In its contemporary state, the theory of climate does not enable us to obtain sufficiently reliable estimates and, therefore, the possibility of more serious global climatic consequences of nuclear explosions as also the irreversibility of the resulting changes in climate cannot be ruled out.

5. THE SUN-STRATOSPHERE RELATIONSHIP

In the extensively studied problem of sun-earth relationship, there still remains the search for the specific mechanism of the impact of solar activity on the atmosphere [3, 16, 23, 26, 29, 64, 90, 91, 126, 133, 178]. The variability, established from observations of ultraviolet solar radiation, solar proton events, variations of galactic cosmic rays and other factors, brings about changes in the composition, radiation regime and dynamics of the stratosphere, which can be considered as most distinct manifestations (mechanisms) of sun-stratosphere-troposphere relationships. It is important to mention, for example, that on the basis of analysis of measurement data obtained from the meteorological satellites *Nimbus-3* and *Nimbus-4*, variations of the solar constant in the 120–340 nm wave band, chiefly due to two active regions of the sun, were discovered.

Simon [156] made a survey of data on extraterrestrial influx of solar radiation in the wavelength interval of 120–400 nm, which is of particular importance for aeronomy in view of the distinct influence of ultraviolet radiations on the photochemistry of the stratosphere, and the problem of solar stratospheric relationships. The principal characteristic of radiation in the interval of 120–175 nm is L_α lines, these being the most intense emission of the chromosphere. Within the limits of measurement errors of $\pm 30\%$, the mean flux of this emission is 3×10^{11} hv cm^{-2} s^{-1}. However, the variability of emission in relation to the 27-day period of rotation of the sun (up to 30%) and the 11-year periodicity (about 40%) should be taken into account.

In [156] different data is provided on the spectral distribution of extraterrestrial flux of solar radiation in the interval of 120–175 nm, indicating the existence of discrepancies amounting to 40% in the interval 150–160 nm, and still more significant discrepancies for wavelengths lower than 140 nm. This may partially be explained by the temporal variability of the intensities of chromospheric emission lines. In the regions of wavelength above 160 nm, discrepancies do not exceed 30%. The discrepancies are also significant (up to 50%) in the interval of 175–240 nm, a range that has not been sufficiently studied. Roughly the same situation prevails in the interval of 240–400 nm.

The insufficient accuracy of measurement of extraterrestrial insolation renders unreliable any data in regard to the dependence of the spectral solar constant of solar activity (this relates particularly to the 11-year cycle). Therefore, improving the accuracy of measurements of extraterrestrial ultraviolet solar radiation by at least 5% and carrying out prolonged continuous measurements, are the most important problems.

The results of balloon measurements of total solar radiation [14,15–17] have aroused considerable interest. A detailed analysis of the data from balloon soundings carried out by the faculty of the Leningrad State University (LSU) during 1961–1970, confirmed the assumption that an anomalous reduction in direct solar radiation at altitudes above 26 km, observed during this period, is due to the presence of products of nuclear explosions in the upper stratosphere [16]. A correlation between the dates of nuclear tests and increase in the reduction of direct solar radiation in the layers above 30 km was also noticed. The reduction of total flux of direct solar radiation in individual cases constituted 6–8% of the extra-terrestrial values (S_0). Apart from this, in [16] it was also stated that during the years 1968–1970, an additional reduction of direct solar radiant flux with a regulated nature and constituting about 1.5% toward 1971, was observed in the upper stratosphere. This additional reduction has been explained on grounds of attenuation of direct solar radiation by stratospheric aerosol.

In recent years, a number of papers have appeared, in which computations of the amount of nitrogen oxides (NO_x) generated in hydrogen bomb explosions of various capacities, have been carried out. Apart from this, it has been explained that in the case of bombs with over one-mega-ton capacity, the products of explosion near the earth's surface are released directly into the stratosphere. For capacities of 30–50 megatons, the fireball is balanced at altitudes of 30–45 km [38]. From approximate estimates, the number of NO_x molecules on the average reaches 10^{32} for a one megaton explosion. According to maximum estimates, the number of NO_x molecules is 1.5×10^{32}. In [87] it is indicated that for explosions in the upper atmosphere, the number of NO_2 generating molecules may increase by 2 orders of magnitude. From available estimates [52], the cumulative capacity of explosions during the period 1961–62 was equivalent to about 340 megatons of TNT. As a result of this, about 1.3–1.7 megatons of NO_x or 5×10^{34} molecules have been released in the upper stratosphere at an altitude of 20–50 km.

183   Computations carried out in [16, 23] showed that the number of NO_2 molecules in a vertical column of a unit cross section, piercing through the semiglobal circular cloud formed as a result of nuclear tests in the middle and higher latitudes of the northern hemisphere should reach 1.6×10^{17}, so as to create an additional reduction of total flux of direct solar radiation equal to 2.6%. According to data mentioned in the said references, the actual number of molecules is about 3.5×10^{16}. If we take into account the fact that tests were conducted even in preceding years (1954–1958), and consider the increase in effectiveness of explosion in the stratosphere (from the quantity of nitrogen oxides produced) in relation to the effectiveness of explosions near the earth's surface, the difference in estimates obtained above would decrease to a minimum.

The proportion of NO_2 molecules, in the examined quantities, from a vertical column of the atmosphere at 50°N in the second half of 1962, confirms the simultaneous data of balloon filter measurements of spectral fluxes of direct solar radiation (in the 305–370 nm band). Hence the conclusion can be drawn regarding the global dimensions of the impact of nuclear tests during 1961–1962, confirming the significant attenuation of direct solar radiation (2.6% or more) and the duration of the effect (1962–1968).

The rate of decrease in excess attenuation of direct solar radiation in the upper stratosphere of temperate latitudes till 1968 was 0.4% per year. During 1962–1963, the stable attenuation of solar radiation in the troposphere and on the earth's surface, at temperate and higher latitudes of the northern hemisphere, was reflected in synoptic processes and climatic conditions. Such consequences of anthropogenic factors lead to the disruption of normal conditions of circulation in the stratosphere and troposphere, i.e. to a perturbation in climate.

A decrease in the absorption of direct solar radiation with time should not be related only to the natural sink of nitrogen dioxide in the troposphere of the northern hemisphere, but also to its transfer toward the upper stratosphere in the southern hemisphere, where, during the period from 1963 to 1966 a highly anomalous attenuation of direct solar radiation occurred. This is mainly based on the data of transport of products of nuclear explosions from higher latitudes of the northern hemisphere toward the equator and then into the southern hemisphere. A comparison between experimental and computational data only confirms the fact that in the southern hemisphere, an overlapping of nuclear and volcanic impacts has occurred. Results of rocket spectral measurements ($\lambda = 380.0$ nm) from the Woomera Station (30°S, 1965) revealed that at 20–40 km altitudes there is an attenuation of components having the same optical thickness as a Rayleigh atmosphere at 28 km, which increased two-and-a-half times at 22 km. Such a loss cannot be considered to arise only on account of the aerosol component. Estimates of the absorption of direct solar radiation at $\lambda = 380.0$ nm, where the maximum absorption of nitrogen dioxide is observed, show that the column of a unit cross section of the atmosphere should contain 7×10^{16} molecules of nitrogen dioxide above 22 km. Such a surplus of nitrogen dioxide would ensure an absorption equal to 1% of S_0 beyond 22 km in the vertical column.

In [14–16] the processing of balloon data for 1968, in accordance with the model presentation of the transparency of upper layers of the stratosphere has been carried out. Therefore, all model and extrapolated data reflects the features of variability of transparency relating to upper layers of the stratosphere as compared with the conditions of transparency in 1968. Such data should be considered as data on the meteorological solar constant ($S_{0, m}$), which is related to the absolute value of the astronomical solar constant $S_{0, ast} = 1371$ W/m² determined from the balloon data of 1967–1968 and undergoing no noticeable variation in relation to solar activity. However, the meteorological solar constant bears a similar relationship, coinciding principally with those proposed earlier for $S_{0, ast}$ [14]. The concept of a meteorological solar constant $S_{0, m}$ which according to accepted views should characterize the 'subzonal' value of $S_{0, ast}$ determined by direct solar radiation flux at the level of the tropopause. should

184

also be modified. It would be worthwhile to consider $S_{0, m}$ as characteristic of this impact of excess stratospheric NO_x (found above 26 km) on
$S_{0, ast}$.

Analysis of the results of winter balloon observations of 1966–1970
enabled us to establish that the attenuation of solar radiation during
winter (up to 1.2%) appears to be a consequence less anthropogenic than
due to the effect of the sun, and seasonal effects in the absorption of
direct solar radiation. In these years, solar activity attained a maximum
value; this is manifested in frequent and stronger influxes of high-energy
solar protons, exerting a significant effect on the processes of ionization
and dissociation of molecules of nitrogen, oxygen, ozone and other components of the upper and middle stratosphere. According to [64], an
intense influx of solar cosmic rays leads to the generation of the same
number of NO molecules at high latitudes (above 60°) in the upper stratosphere, as in the case of an explosion of a nuclear device of 50 megatons
or more. The possible decrease in ozone concentration associated with
such an explosion at an altitude of 45 km would be 25%, and the concentration of nitric oxide would go up 2–3 times, which would lead to
185 an increase in the number of NO molecules in the vertical column by
6×10^{15}. Such an increase exceeds approximately 4 times the contribution made by galactic cosmic rays in the same latitudinal zone.

An important property of galactic cosmic rays is their high penetra-
tive power. Therefore, a major absorption of galactic cosmic rays occurs
at an altitude of 9–16 km, where a localized minimum of water-vapor
and ozone concentration, and a localized maximum of nitrogen-dioxide
concentration are located. The effects of solar and galactic cosmic rays
on the composition of the stratosphere are resolved with respect to time
(various periods of the solar cycle), and also altitude: the maximum of
solar cosmic-ray absorption lies in the altitude range of 35–40 km, where
the proportion of nitrogen-dioxide mixture is maximum. The combined
impacts of solar and galactic cosmic rays, apparently, determine the dependence of $S_{0, m}$ on solar activity obtained after excluding the effect of
anthropogenic influences during 1961–1967 [15, 16, 23]. The maximum,
which is localized in the region of smaller Wolf numbers ($R_z = 40$–70) and
the amplitude of variation of which does not exceed 1.2% as compared
with its position on the initial curve of S_0 has slightly shifted.

The decrease in $S_{0, m}$ (in the region where the Wolf number exceeds
the value 100) is related to the influence of high-energy solar protons on
the chemical composition of the middle and upper stratosphere, which is
reflected mainly in the increase of NO_x at altitudes of 35–45 km in the
temperate and, especially, higher latitudes. Decrease of $S_{0, m}$ in the case
of transition into the region of small values of R_z is caused by galactic
radiation, the intensity of which is maximum at the minimum of solar

activity. Hence, the contribution of galactic cosmic rays to values of the solar constant, measured by means of a balloon, should be smaller than that of solar cosmic rays, since galactic cosmic rays are mainly absorbed in the lower stratosphere.

Thus the effect of solar activity on the atmospheric processes is realized mainly through variations in its chemical composition (ionization and dissociation of molecules of nitrogen and oxygen, formation of nitric oxide and, thereafter, nitrogen-dioxide with subsequent active participation of nitrogen-dioxide and transformation of solar radiation into heat (330–550 nm band). The 'modulation' of the composition of the stratosphere and corresponding radiant heat influx caused by galactic and solar cosmic rays appears to be the main mechanism of sun-troposphere relationships, in which the stratosphere plays the role of a controlling link.

Ruderman and Chamberlain [149] observed the formation of nitric oxide in the stratosphere under the influence of solar cosmic rays, which leads to the destruction of traces of oxygen. SCR are modeled by a sectoral structure of the interplanetary magnetic field, which is apparently related with the active regions on the sun. Theobald et al. [168] undertook the analysis of the combined effect of these factors of solar activity on the middle atmosphere (stratosphere and mesosphere).

For the 190–310 nm wave band, the following approximation of amplitude of relative variation of the spectral solar constant depending on the wavelength has been adopted:

$$\Delta S/S = 0.00107 \exp (669.6/\lambda),$$

which corresponds to a 0.04% variation of the total solar constant. The amplitude of variation for solar cosmic ray flux has been taken from data relating to the period of a relatively smaller solar flare in September 1966. Estimation of the influence of solar activity on the odd oxygen and vertical profile of temperature were carried out by the application of a simplified photochemical model, which takes into account the simple 'oxygen' chemistry, modeled by the formation of nitric oxide under the influence of solar cosmic rays. The effect of atmospheric movements is not taken into account but the presence of a photochemical equilibrium is assumed. Sinusoidal variations of the solar constant with periods of 14 and 18 days, as well as 'flares' repeating every 14 days, have been assumed.

Computations of the vertical profile of temperature in the 10–95 km layer in April at the equator by assuming sinusoidal (14 day) perturbations of the ultraviolet solar constant enabled us to reveal a maximum perturbation of temperature with an amplitude of about ± 1 °C in the middle stratosphere. The amplitude decreases above 40 km (approximately up to ± 0.6 °C at the level of 50 km), but close to the level of 70 km there exists a secondary maximum with an amplitude of 0.5 °C. Above

85 km the amplitude is almost independent of altitude and is much smaller in magnitude (about $\pm 0.1\,°C$) than the natural variability of temperature.

The simultaneous consideration of variations of the solar constant and cosmic radiation (computations of temperature profiles have been carried out for points at 45°N, 225°E in April) gave similar results for the stratosphere at a smaller maximum amplitude (about $\pm 0.6\,°C$) which is due to dependence of the ultraviolet solar radiation flux on latitude. Significantly different results have been obtained for the mesosphere, where the temperature varies in the range of about $\pm 0.8\,°C$ and with a phase lag relative to the variability of stratospheric temperature. This should be explained by an increase of nitric oxide concentration in the mesosphere with an increasing intensity of cosmic ray flux which causes a decrease of ozone concentration and temperature. The maximum change in ozone concentration is observed at a height of 80 km and attains 10%.

187 Consideration of more realistic variations of the ultraviolet solar constant, as compared to a sinusoidal, for points at 45°N, 90°E in April showed that warming or cooling of the stratosphere with an amplitude of 0.4 °C results from the variability of the solar constant, and thermal regime of the mesosphere, where the temperature fluctuation within $\pm 0.8\,°C$ is determined by the variation in cosmic radiation. For more reliable estimates of the impact of solar activity on the middle atmosphere, further investigations on the variability of the spectral solar constant are essential. Further analysis of the role of the middle atmosphere as a connecting link in transmitting the impact of solar activity to the troposphere is also of great significance.

Diffusion processes ensure the transfer of hydrogen from the troposphere to the middle atmosphere (stratosphere and mesosphere) in the form of methane and water vapor. After chemical- and photodissociation of these components, hydrogen (in combined or atomic form) enters into active chemical reactions, leading to various types of influences on the stratospheric ozone. The effect of HO_x on ozone can be either positive or negative depending on the altitude, and a reliable determination of the total global effect on the middle atmosphere is not yet possible.

In the upper mesosphere, atomic hydrogen enters into a reaction with ozone as follows:

$$H + O_3 \rightarrow OH^{\bullet} + O_2^{\bullet},$$

which is the main source of emission of hydroxyl ions, constituting one of the important components responsible for night glow. In view of the fact that the middle atmosphere serves as a connecting link between the thermosphere, where solar activity is clearly manifest, and the tropo-

178

sphere (the influence of solar activity on this is disputable), Weill and Christophe [177] undertook the analysis of observational data on the following two interrelated characteristics, which depend on the hydrogen content in the middle atmosphere: the intensity of hydroxyl emissions at the level of the mesopause, and concentration of water vapor above the tropopause. The latter has been taken from the data of measurements of dew point with the help of a hygrometer during the period January 1964 to November 1975, near Washington, D.C. (data of monthly measurements for the 70 mb level has been considered). The intensity of hydroxyl emissions has been taken from observational data obtained at Haute Provence Observatory during the period February 1954 to December 1975. The data for the wave interval centered at 634 ± 19 nm, where the emissions in the (9–3) OH band with small contributions of the (5–0) and (6–1) OH bands and the continuum appear, have been studied.

The study of the long-term variability and autocorrelation analysis of data on hydroxyl emission confirms an earlier conclusion about the presence of annual and semi-annual trends as the main periodicities. The annual trend of water-vapor concentration at the 70 mb level is less distinctly manifest. While comparing the secular varibilities of intensity of hydroxyl emissions, water-vapor concentration and a solar radiation flux at a frequency of 2800 MHz, which serves as an index of solar activity, the presence of a certain correlation (with a phase lag) with water-vapor concentration has been revealed. Over a period of 8 months, the maximum correlation coefficient of the index of solar activity and concentration of water vapor was 0.38 ± 0.06.

Analysis of the variability of hydroxyl emissions showed that a correlation of the product of concentrations $[O_3] \cdot [H]$ and the 11-year cycle of solar activity is evident at an altitude of 85 km, and the ratio of maximum and minimum values is equal to 1.6. The results thus obtained enable us to assume that there exists a semi-annual phase between variations of atmospheric composition observed at the level of the mesopause and in the stratosphere. This assumption, being of significance for the problem of impact of solar activity on the lower atmosphere, requires verification on the basis of data over prolonged observations. The most probable influence may include the fact that the solar cycle modulates the physical characteristics of vertical movement at the level of the mesopause and in the mesosphere, resulting in variations of such optically active components of the stratosphere as water vapor and ozone, which lead to small variations in the radiation budget at the level of the tropopause and below it.

Chapter III

The Effect of Aerosol on Radiative Transfer and Climate

189   The reasons for assuming the effect of naturally occurring aerosol on climate have been recognized for quite some time. This becomes more evident during a period of volcanic eruptions, when a decrease in air temperature is observed. On the other hand, the absence of severe eruptions over a period of several decades was considered as the main reason for a warming in the first half of the present century [2]. The effect of eruptions on air temperature was most distinctly observed in the stratosphere. For example, after the eruption of Mount Agung (Bali Island) on March 17, 1963, the temperature in the stratosphere of the tropics and subtropics of the southern hemisphere had increased by about 5 °C and this increase was noticeable even several years after the eruption. It should be mentioned here that, although the effect of this eruption on the attenuation of solar radiation was clearly recorded even in the temperate latitudes of the northern hemisphere, i.e. the effect was global, no noticeable changes of temperature in the layers lower than 300 mb were found in the tropical zone of the southern hemisphere. At the same time, a precise impact of warming in the stratosphere of higher latitudes was also not observed. Therefore, the conclusion may be drawn that an increase in the temperature of the stratosphere at lower latitudes has been caused directly through absorption of solar radiation by aerosol. As far as the attempts to attribute the cooling after 1940 to the effect of intensified eruption in this period are concerned, they should only be considered as hypothetical.

In recent years, there is increasing evidence of the existence and influence of large-scale streams of dust and sand, developed in the desert regions and moving across long distances, on the thermal regime of the atmosphere. This relates especially to the Sahara Desert and the transport of dust across the Atlantic Ocean. No doubt, the influence of the aerosol factor on climate requires careful investigation, and this was particularly examined during the implementation of the GATE project (1974).

Presently information regarding the considerable growth in the intensity of industrial aerosols being released into the atmosphere is being accumulated. It is stated, for example, that during the past decade the 190 aerosol content in the rural areas of the USA has increased by 20–30 %. At the same time, analysis of measurement data on the transparency of the atmosphere from the beginning of the twentieth century led to the conclusion that if we were to exclude the sporadic variations caused by volcanic eruptions, no noticeable global tendencies of changes in transparency over the last fifty years have been observed. Interesting information regarding increase in the haziness of the atmosphere in urban zones and industrial regions has been obtained by analyzing photographs and television pictures of the earth taken from outer space.

In this, the problem of estimation of individual contributions made by aerosols of industrial and natural origins is very important. The ratio between aerosols of different origins varies substantially in space and depends on the distribution of the sources of these aerosols. According to global estimates the proportion of industrial aerosol is 6%. This estimate has been obtained under the assumption that the characteristic period of deposition of particles is 30 days. The adequacy of this assumption, however, is still not proved.

Undoubtedly, if we make an attempt to predict the future, it would appear that the number of industrial sources of pollution will sharply increase (of course, the reliability of extrapolation of the present-day tempo of growth of aerosol pollution is doubtful, since serious steps for bringing down the level of industrial pollution are already being taken). Measurements of electrical conductivity over the Atlantic Ocean from 1910 to the present indicate that the bulk concentration of particles in the northern hemisphere has increased approximately two times whereas the average size of the particles has decreased 1.5 times. The mass concentration of particles has, apparently, increased 1.5 times. According to available data, the annual increase in the industrial aerosol content up to the year 2000 would constitute 0.3–0.4%.

The seriousness of the problem concerning the possible effect of aerosol (natural and, particularly, industrial) on climate has created considerable interest and has given rise to a whole series of investigations devoted to approximate estimations of the effect of aerosol [1–4, 8–10]. However, it may be mentioned that the results of these investigations to a considerable extent, are contradictory and mainly determined by excessive schematic simplification of many models and an inadequacy of basic information in regard to aerosols.

The absence of reliable data on global aerosol and its characteristics (concentration, microstructure, shape of particles, chemical composition, optical parameters) renders unrealistic any effort to obtain a sufficiently reliable description and prediction of the effect of aerosol on climate. The available results, however, enable us to consider individual aspects of this problem, which are of great interest from the viewpoint of formulation of unresolved questions and directions of further investigations with a view to obtaining a reliable solution.

1. TROPOSPHERIC AEROSOL

The discussion on climatic effects of tropospheric aerosol has aroused considerable interest in regard to the problem of radiative transfer in an absorbing and scattering medium [8, 9, 17–20, 25–27, 29, 36, 47–52, 58–73, 78–82, 85, 89–92, 95–100]. Based on the application of various methods of calculation of radiation fluxes, computations of the effect of the albedo of the earth's surface-atmosphere system, of aerosol on the absorption of radiation by the atmosphere, and of climatic tendencies (warming or cooling) were undertaken. Insofar as these problems have been dealt with in recently published monographs [1–3, 10, 15, 59] and a large number of papers, they will not be discussed in detail here. It should be particularly emphasized in the first place that the radiational properties of aerosol are complicated functions of aerosol characteristics (concentration, microstructure, complex index of refraction) and a number of other parameters (albedo of the underlying surface, altitude of the sun, etc.).

The data of direct measurement, and application of methods of cosmic remote-sensing, progressively demonstrate that till now the role of aerosol as a factor affecting the fluctuations of weather and climate has obviously been underestimated. This became especially clear, where intense absorption of shortwave radiation by aerosol was discovered. The significant effect of aerosol on the transfer of thermal radiation is also evident.

The necessity of investigating atmospheric aerosol, of natural or anthropogenic origin, as one of the factors of climatic variations determines the importance of observations of its properties without which it is impossible to establish the role of anthropogenic aerosol. Observational data at the South Pole enabled us to establish, that many chemical elements constituting anthropogenic aerosol (aluminum, iron, titanium, manganese, lead, vanadium, calcium, lanthanum, scandium, thorium, cesium, samarium, tantalum, hafnium, En*, copper) are the products of weathering of rocks, and that the contribution of the oceans is determined mainly by the concentrations of sodium, as also of magnesium, calcium,

*Probably Eu (europium)—Translator.

potassium, chlorine and bromine. Concentrations of such elements, as tungsten, indium, gold, antimony, iodine, zinc, copper, arsenic, cadmium, selenium and large amounts of bromine appeared to be $10-10^5$ times
192 higher than the expected concentrations of these elements in rocks or oceanic salts. The relative mass concentration of 'anomalously enriched elements' (AEE) is 5-10%. Similar data has been obtained from observations in the northern parts of the Atlantic Ocean.

The main bulk of Antarctic aerosol consists of sulfates having, apparently, a stratospheric origin, because of AEE, as well as their oxides and halides which are fairly volatile. It may therefore be assumed that an enrichment has been caused by natural or anthropogenic processes taking place at elevated temperatures. From this point of view, volcanoes can be one of the important sources of AEE. In this connection Lepel et al. [63] undertook a chemical analysis of filter samples of volcanic aerosol, taken at the earth's surface and in the free atmosphere (from an aircraft) in the region of Mount Augustine (Alaska).

The samples were collected during four reconnaissance flights between February 1 and 21, 1976, when a series of eruptions occurred, having originally started on January 22, 1976. Ground samples of volcanic ash were collected at various points over the territory of Alaska and along the shores of Augustine Island (at a distance of 10 km from the crater). For comparison with aircraft data, the results of analysis of ash from the latter point have been used. The application of neutron activation analysis and mass absorption spectrometry enabled us to determine concentrations of 38 elements.

The results, thus obtained, demonstrate that concentrations of volatile elements in the free atmosphere (zinc, copper, antimony, gold, lead, arsenic, cadmium, chlorine, bromine, selenium, mercury, sulfur) vary between $10-10^3$ times as compared with concentrations of these elements in volcanic ash, whereas in the case of nonvolatile components such an enrichment is not observed. The results are very similar to those obtained for other volcanic effusions and background aerosol in remote regions of the globe. For samples obtained at an early stage of the eruption cycle, a higher content of volatile elements is characteristic as compared to samples collected at a later stage. This indicates significant changes in the composition of effusions in the course of the eruption cycle. Such a variability exists, depending on the type and location of the eruption. Investigations on the chemical composition of products of volcanic eruptions are still at the initial stage. Furthermore, regular observations with the aim of obtaining data characterizing the process of fractionation, which takes place as the volcanic flame ages, are necessary. Observations relating to various volcanoes alone would not enable us to assess the contribution of volcanoes in the formation of global background aero-

193 sol. However, it may be assumed that volcanic effusions appear to be the most probable source of such evanescent elements as zinc, copper, gold, arsenic, lead, cadmium, chlorine, bromine, antimony, on a global scale.

1.1. Absorption of radiation by aerosol. Since the early fifties, when Kastrov first observed that the measured absorption of shortwave radiation in the free atmosphere significantly exceeded the calculated molecular absorption, many investigations have confirmed this result. The aircraft measurements of spectral shortwave radiant fluxes first carried out during the CAENEX expedition* [3, 15, 60] and later confirmed by data of GATE [16], appeared to provide a direct proof of the existence of considerable (but highly fluctuating) aerosol absorption. The data of these measurements demonstrate, that the earlier detected 'residual' absorption of shortwave radiation is, in reality, an aerosol absorption possessing significant selectivity. Its possible variations are determined by fluctuations of aerosol concentration in the atmosphere as well as a variability in the chemical composition and consequently the optical characteristics of aerosol.

The determination of radiant heat influx on the basis of measurement data of radiation fluxes is a complicated problem, since in this the necessity of computing second-order differences emerges. It is natural that such a problem can be solved only by the application of hypersensitive measurements. In order to obtain more reliable results, the values of radiation influx for a thick layer of the atmosphere were determined. The search for methods of data processing to determine vertical profiles of the effective flux of shortwave radiation $F_\lambda^\uparrow - F_\lambda^\downarrow$ albedo $A_\lambda = F_\lambda^\uparrow / F_\lambda^\downarrow$ and radiation influx $b_\lambda = \Delta (F_\lambda^\uparrow - F_\lambda^\downarrow) \Delta z$, led to the use of a statistical method described in the review article [10].

Vertical profiles of b_λ beyond the molecular absorption bands indicate a decrease in radiant heat influx with increasing altitude caused by a decrease of aerosol concentration. In the Chappuis ozone bands, b_λ is practically unchanged with altitude, because a decrease of aerosol concentration is compensated by an increase of ozone content in the layer $\Delta p = 100$ mb. In all other absorption bands, b_λ decreases with altitude. In oxygen bands the heat influx remains uniform and does not increase intensely while approaching the earth's surface. But in water-vapor bands a very sharp increase of flux in the lower layers of the atmosphere is observed, which may be explained by a concentration of water vapor in
194 the lower layers of the atmosphere, especially in the case of observations over desert regions.

An analysis of results of aircraft measurements on vertical profiles of components of the radiation budget during daytime, in the region of the

*CAENEX (The Complete Atmospheric Energetics Experiment) —Translator.

GATE operation, showed that in conditions of a cloudless and dust-free atmosphere, the presence of a maximum in the lower one-kilometer layer and a subsequent drop up to an altitude of 7–8 km (attained in the aircraft sounding) is characteristic for vertical profiles of radiant heat influx. Moreover, these peculiarities are determined by the formation of profiles of radiant heat influx under the influence of shortwave radiation (cooling caused by long-wave radiation is insignificant, about 7 W/m^2 km, and changes only slightly with altitude).

During the days of dust outbreaks over the African continent, when the Saharan aerosol layer (SAL) is formed, a sharp maximum of solar radiation absorption is observed (about 700 W/m^2 km) in the layer of maximum concentration of aerosol (3–4.5 km). Within the limits of the SAL a warming due to long-wave radiation takes place. Above the SAL, the radiant heat influx is close to zero as a consequence of mutual compensation of the shortwave and long-wave components of radiant heat influx. The data on spectral measurement of upward and downward fluxes of shortwave radiation in the 0.4–0.9 μm wave band reveal a considerable selectivity of absorption, which is exclusively caused by aerosol in the 0.4–0.6 μm band. The spectral relationship of absorption also affects the variability of the complex index of refraction with wavelength, having a maximum at 0.41–0.42 μm. During periods of dust outbreaks, the absorption of shortwave radiation by the 0.5–6.3 km layer attains values of 20%, and exceeds roughly by two times the amount of molecular absorption. On the other hand, in conditions of a dust-free and cloudless atmosphere, it decreases to 5–7%.

The layer of dense stratocumulus clouds of 200–500 m thickness (at the altitude of the upper boundary, 900–1200 m) acts as a sink of heat flux with a capacity of up to 70–170 W/m^2 km. In the 1–1.5 km layer above the clouds, radiant heating is observed and the 3–4.5 km layer is close to radiative equilibrium. Measurements of the microstructure of clouds revealed that clouds in the eastern sector of the tropical Atlantic are characterized by an increased concentration of coarse particles (size up to 1 mm) as compared with clouds of continental origin. For this reason the optical thickness of clouds above the ocean is considerably lower.

For an interpretation of data obtained under the CAENEX program, their comparison with the results of numerical experiments for actual conditions of observations is important. A joint Soviet-American program was launched with the objective of analyzing the data of comprehensive observations of the spectral fluxes of shortwave radiation and properties of atmospheric aerosol carried out on October 25, 1970, in the Karakum deserts by the CAENEX-70 expedition. The results of this program are presented in [1]. The analysis has been carried out by comparing observed values of spectral fluxes and influxes of shortwave radiation with computed results for specific conditions of observations, primarily with the aim of assessing the effect of aerosol on the transfer of shortwave radiation in an actual atmosphere.

The day the observations were taken was characterized by a hazy atmosphere after an intense dust storm. On the basis of data of aircraft measurements of upward and downward spectral fluxes of shortwave radiation in the 0.4–0.9 μm wave band, the spectral distribution of fluxes in the 950–350 mb layer have been calculated by modulation and interpolation with altitudinal steps of 100 mb. Analysis of the microstructure and composition of aerosol showed that the chemical composition of particles remains sufficiently homogeneous, and it was possible to assess (by electron microscopic investigations) submicronic fractions of individual compact spherical particles, which were identified as oxides of iron (hematite). The radius of the particles was about 0.1 μm.

Calculations of the coefficient of absorption, scattering and attenuation, and subsequent computations of radiation fluxes have been carried out under the assumption of a spherical shape of the particles, a homogeneity of their chemical composition and an independence of the complex index of refraction from wavelength. Aerosol was considered to be composed of a polydisperse fraction and an individual submicronic fraction of hematite particles. The complex index of refraction has been taken as 1.65–0.005i. The measured microstructure of aerosol was extrapolated to a radius in the range of 0.03–10 μm.

Calculations of vertical profiles of coefficients of absorption, scattering, and attenuation confirm an increase of these coefficients with altitude, which is caused by variation of the microstructure of aerosol in relation to height. The calculations, carried out for various models of aerosol were not, even in a single case, found to be in complete agreement with observational data of spectral fluxes of shortwave radiation, but definitely showed (in agreement with the observations) a significant contribution to absorption because of small amounts of impurities consisting of highly absorbent matter present in aerosol.

196 Varying the initial parameters led to the conclusion, that computed coefficients of absorption, scattering, and attenuation are very sensitive to the various components of aerosol. The shortwave radiation fluxes have been calculated by the method of spherical harmonics with the application of Eddington's delta-approximation. The calculations are based on a ten-level model of the atmosphere. It appeared that significant differences between the measured and calculated values of upward radiation flux have been caused mainly by a significant systematic underestimation of measured values.

The introduction of corresponding corrections in measurement data and the use of an aerosol model in which the presence of individual fractions of absorbing particles, and of hematite 'immersed' in coarse particles of aerosol was considered to have provided the best agreement with observational data. Even in conditions of an extremely hazy atmosphere, upward radiation fluxes closely depend on the albedo of the underlying surface. In subsequent implementation of the comprehensive program of aerosol and radiation observations, it is essential to foresee the possibility of obtaining considerably more complete and reliable information on the properties of aerosol (chemical composition and its relationship to size of the particle, structure of individual particles, etc.) as well as an increased accuracy of spectral measurements of shortwave radiation fluxes.

Braslau and Dave [26, 27, 39] carried out calculations of vertical profiles of radiant heat influx due to absorption of solar radiation for six models of atmosphere from the pure Rayleigh model to that containing absorbent aerosol and a dense stratus-type cloudiness in the 3–4 km layer. The calculations of influxes have been carried out for a layer of the atmosphere of 1 km thickness in the altitude range 0–30 km. Analysis of the variability of vertical profiles of radiant heat influx for various models of the atmosphere, depending upon the zenith distance of the sun and albedo of the underlying surface, showed for a zenith angle of the sun $\theta = 0°$ (up to 30%), an increase of absorbed solar radiation near the earth's surface. For a higher aerosol content, the radiant heating is doubled as compared with heating in the condition of a dust-free atmosphere, but in the 5–15 km layer, where the aerosol content does not increase, the radiant heating remains almost unchanged. In Young's layer, the aerosol absorption constitutes 17% of molecular absorption.

When clouds are present, most severe changes in radiant warming take place near their boundaries (increase above, and decrease below the clouds). Within the layer of cloudiness, the absorbed radiation increases roughly by 15%, which is caused by multiple scattering. When $\theta = 40°$, the vertical profiles of radiant heating are similar to the profiles for $\theta = 0°$, but the rate of warming decreases, since incident flux decays in proportion to $\cos \theta$. More significant seems to be, however, the increase in the rate of warming above the clouds in view of the increase in albedo. For $\theta = 80°$, the maximum heating takes place above the clouds as a consequence of a sharp increase in their albedo. The major peculiarities of vertical profiles of radiant heat influx are determined by the effect of molecular absorption by water vapor and ozone. The presence of purely dispersive aerosol provides a small contribution to radiant heating.

The exact solution of the problem concerning the effect of aerosol on radiative transfer in the atmosphere requires a consideration of the relationship between the optical properties of aerosol and humidity of the air. Zhunkowski and Liou [100] studied the relationship between the optical properties of aerosol in the visible region of the spectrum and humidity (using relative humidity in the range of 30–70%) for the conditions of a highly polluted atmosphere on the basis of application of three models of atmosphere, characterizing fairly humid and dry air, as well as typical conditions of humidity at temperate latitudes. The microstructure of aerosol has been taken in accordance with Young's formula [100], and the vertical profiles of concentration correspond to the model of McClatchey et al. The aerosol concentration near the earth's surface is equivalent to a horizontal visibility of 5 km for a relative humidity of 50%.

In [100], calculations have been carried out (by the method discrete ordinates) for transparency, absorption, radiant heating and albedo of the underlying surface-atmosphere system at 45°N latitude in mid-February for an albedo of the underlying surface equal to 15%. The relationship between the radii of aerosol particles and relative humidity has been considered in accordance with the Henel data. The effect of humidity on the scattering index was not taken into account.

The results of calculations of absorption and vertical profiles of radiant warming on the 0.55 μm wavelength showed that the effect of humidity in these cases is insignificant and the profile of heating is similar to the vertical profile of aerosol concentration. The transparency of the atmosphere however undergoes considerable variations (up to 9% at midday), depending on humidity.

At all zenith angles there is an increase in the albedo of the system with increase in relative humidity. The maximum difference in the values of albedo at 09 : 00 h is 5%.

In [100] estimates of mean global values have been obtained with the aim of analyzing possible climatic effects. Mean global values show that the effect of humidity on absorption and radiant warming is practically insignificant, but variations of transparency and albedo of the system because of humidity are considerable (the increase in albedo with increasing relative humidity goes up to 20%). This means that such variations should be taken into account for the study of the impact of aerosol on climate.

Halpern and Coulson [47, 48] carried out interesting calculations showing the effect of various characteristics of aerosol (microstructure, complex index of refraction, and vertical distribution of concentration) on the outflow and inflow of shortwave radiation in the boundary layer of the atmosphere. The basis of these calculations lies in the model of a cloudless atmosphere characteristic for the conditions of summer at temperate latitudes with a boundary layer of 2 km thickness. The vertical profiles of atmospheric pressure, temperature, concentration of water vapor and ozone have been taken from the data for a standard atmosphere developed by McClatchey et al. The total contents of water vapor and ozone are 1.826 g/cm² and 0.005 atm cm respectively.

With the objective of modeling the impact of anthropogenic aerosol on the transfer of radiation, it is assumed that the total content of particles in a vertical column of the atmospheric boundary layer increases from 1.646×10^7 (model C, proposed by Braslau and Dave) up to 1.886×10^8 cm^{-2}, and three different vertical profiles of concentration have been considered. The calculations were undertaken for two models of microstructure of aerosol: the haze model L of Deirmendjian and the Young distribution. Since in both cases, identical masses of aerosol in a vertical column of the atmosphere are taken, the total number of particles equal to 1.886×10^8 cm^{-2} in the case of haze L increases to 1.368×10^9 cm^{-2} for Young's aerosol. The complex index of refraction varies within the following limits:

Index of refraction	Aerosol absorption
1.50–0.00i	Absent
1.50–0.01i	Moderate
1.50–0.10i	Intense
1.66–0.00i	Absent
1.66–0.25i	Very intense
1.80–0.50i	Very intense

Analysis of the results of computations of vertical profiles of shortwave radiation fluxes in the boundary layer of 80 wavelengths in the interval 0.30–2.5 μm showed that, even in the case of moderately absorbed aerosol, the amount of aerosol absorption exceeds the molecular absorption by approximately four times. The same value will be ten times greater, if the complex index of refraction is equal to 1.50–0.10i. The haze L causes more intense absorption than Young's aerosol (for an identical index of absorption). The distribution of absorbed radiation along altitude significantly depends on the peculiarities of the vertical profile of aerosol concentration and solar altitude. The results, thus obtained, demonstrate the necessity of reliable complex measurements of vertical profiles of radiation fluxes, the concentration and microstructure of aerosol, as well as the complex index of refraction of the particles.

Liou et al. [65] carried out numerical modeling of shortwave radiation transfer with the objective of calculating vertical profiles of radiant heating due to absorption of solar radiation for the following four models of the atmosphere: 1) molecular; 2) molecular + aerosol; 3) molecular + cloud; 4) molecular + aerosol + cloud. As compared with the calculations carried out earlier, in fresh calculations, the vertical inhomogeneity of a plane-parallel atmosphere and molecular absorption by the scattering layers have been taken into account. The vertical profiles of aerosol concentration correspond to a clear atmosphere with a horizontal visibility of 23 km. The complex index of refraction was used for hygroscopic dust particles depending on their wavelengths. The microstructure of aerosol, which is independent of altitude has been given by an exponential distribution.

The lower boundary of clouds (cumulus clouds during fine weather) has been assumed at an altitude of 1.7 km, and the thickness of the layer has been taken as equal to 0.45 km. The vertical profiles of water vapor and ozone concentrations correspond to the model of the tropical atmosphere (total moisture content equal to 5.2 g/cm^2). Calculations, using the method of discrete ordinates, were carried out by dividing the entire spectrum of solar radiation into nine intervals, centered at wavelengths from 0.3–3.2 μm. This is convenient from the viewpoint of consideration of molecular absorption by water vapor, ozone, carbon dioxide and oxygen. The entire vertical thickness of the atmosphere has been divided into 12 layers of 3 km thickness except for the three lower layers, in one of which cloudiness prevailed. The optical properties of each layer are determined by assuming an optical thickness, albedo of single scattering and the coefficients.

Results of computations showed that, in the presence of a layer of cumulus clouds in the lower troposphere, radiant heating in the clouds attains 12 K/day, if the sun is at the zenith. Absorption by water vapor and droplets provide a major contribution to heating. In the conditions of a clear atmosphere, the effect of aerosol is clearly manifest. For a surface albedo of 0.1, the maximum radiant heating in a purely molecular atmosphere takes place at an altitude of about 2 km and is equal to 3.4 K/day (the main contribution is from water vapor). Additional aerosol 200 leads to a most significant (approximately 0.5 K/day) additional heating near the earth's surface. Approximately similar heating is caused by an increase in surface albedo from 0.1 to 0.4.

In the presence of clouds the contribution of aerosol in radiant heating is small. The rate of radiant heating sharply decreases with an increase in the zenith distance of the sun. For a lower sun, the effect of clouds on radiant heating is insignificant. The presence of a layer of clouds in the troposphere caused additional heating on account of absorption of the upward radiant flux by ozone in the lower stratosphere. The effect of the underlying surface albedo (values of albedo taken as 0.1 and 0.4) appears only in the case of an absence of clouds. A noticeable contribution to the total absorption of radiation by the layer of the atmosphere arises because of absorption, caused by ozone and oxygen. Neglect of these components leads to an underestimation of the total absorption by 7 and

13% respectively, for conditions of a clear and cloudless atmosphere.

In connection with the discussion of a possible impact of aerosol on climate, the investigations on the relationship between absorption and scattering of solar radiation by aerosol have acquired significant importance. It is found that when scattering by aerosol predominates, an increase in dust-content of the atmosphere should result in an increase in the global albedo and, consequently, a climatic cooling. Paltridge and Platt [75] analyzed the relationship between absorbed and scattered radiations due to aerosol on the basis of data obtained from aircraft measurements of total shortwave and long-wave radiation fluxes as well as radiation on a wavelength of 11 μm in the case of a dense layer of aerosol in the atmosphere. Measurements have been carried out above the sea in November-December, 1971, off the shores of New South Wales (Australia).

Analysis of the data of soundings of November 18, 1971, when the vertical visibility was less than 0.5 km and the upper boundary of the hazy layer was located at an altitude of 2–9 km (altitude of the sun during flight varied from 60 to 48°), showed that considerable absorption of shortwave radiation by aerosol takes place only below the upper boundary of the inversion layer located at the 1 km level. Derived at zero zenith distance of the sun, the value of absorbed shortwave radiation was about (6 ± 1) mW/cm².

Scattering of radiation by the entire aerosol layer, determined from variations of the upward flux of shortwave radiation as compared with its value in conditions of a transparent atmosphere, exceeds the total absorption by the layer by almost 1 mW/cm². Hence, if the effect of aerosol on long-wave radiative transfer is neglected (analysis of measurement data did not reveal considerable variations of long-wave radiation fluxes in the aerosol layer), the total effect of increase in dust loading consists in a slight cooling of the earth's surface-atmosphere system in the given case. However, it is important that scattering chiefly occurs above that layer where absorption mainly occurs. In the absorption layer (below 1 km), scattering by aerosol is less than absorption by approximately 4 mW/cm².

1.2. Albedo of the system. One of the most important questions is how the albedo of the earth's surface-atmosphere system and the corresponding absorbed solar radiation vary along with the content and optical properties of aerosol. In [3] and several other works [7, 8, 62–65, 67], the results of numerical modeling carried out with the aim of resolving this question are described in detail. It is not difficult to understand that the variability of the albedo of the system is determined not only by the optical properties of the atmosphere (including aerosol) but also by the albedo of the underlying surface.

Calculations showed that even for a very small albedo value of the underlying surface, but with intensely absorbent aerosol, a 'warming' effect of aerosol can be observed. Therefore, it is clear that the assertions, often encountered regarding the increase of dust content of the atmosphere being accompanied by an increase in the albedo of the system, and consequently, the tendency toward climatic cooling, are justified.

Results of the aforementioned calculations relate to the conditions of a sufficiently realistic model of the atmosphere. Nevertheless, an analysis of the experimental data is of great interest, enabling an explanation of the actually observed situation. Such data was first obtained during the expedition in the CAENEX program [3, 15, 60]. The obtained data confirms the qualitative conclusions of theoretical computations.

Various criteria characterizing the impact of aerosol on climate (warming or cooling) have been proposed by a number of authors [3, 7, 22, 32–35, 46, 51, 62, 79–81, 96]. These criteria are physically easily formulated with the help of the concept of effective albedo A_e of a layer of aerosol particles, as shown in [3]: if A_e is greater than the surface albedo, the appearance of aerosol in the atmosphere leads to a higher albedo of the system, and consequently to a cooling of the planet. If A_e is less than the surface albedo, this appearance of aerosol in the atmosphere leads to a lower albedo of the system and consequently to a warming.

The effect of aerosol on the climate of the earth is illustrated by the data presented in Table III.1. Considering that two-thirds of the earth's surface is covered by seas and oceans, it can be assumed that the appearance of aerosol should lead to a cooling of the earth. However, as more than half of the earth is constantly covered with clouds (i.e. has a higher albedo), it may be considered that for the earth as a whole, an increase of dry aerosol content in the atmosphere should cause a trend toward climatic warming.

The results of calculations, carried out by Wang and Domoto [95], on shortwave radiation (including the albedo of the earth's surface-atmosphere system) show that the impact of warming or cooling depends on the combined effect of cloudiness, aerosol and albedo of the underlying surface on the albedo of the system. In the case of the model of the atmosphere accepted in [95], an increase in haziness of the atmosphere (caused only by aerosol) leads to a decrease in the global albedo of the system (warming) for an albedo of the underlying surface exceeding 0.30, which corresponds to conditions of snow-covered, desert or dry-grass surface. As the mean albedo of the underlying surface is equal to 0.105, it follows that an increase in haziness of the atmosphere leads to a rise in the albedo of the system, i.e. to a cooling of the earth's surface-atmosphere system.

In case of a 'clear' atmosphere or average conditions of cloudiness,

Table III.1. Effective albedo of aerosol atmosphere for various values of complex imaginary part of complex refractive index m and size distribution of particles

Material	m	β^{1}			Effect on climate		
					above clouds and snow ($A=0.70$)	above vegetation, cities, etc. ($A=0.15\text{-}0.20$)	above water ($A=0.8$)
		2	3	4	total		
Water[2]	$1.5+0i$	1.00	1.00	1.00	} Cooling	} Cooling	} Cooling
	$1.5+10^{-4}i$	0.93	0.99	0.98			
	$1.5+5\times10^{-4}i$	0.72	0.93	0.91			
	$1.5+10^{-3}i$	0.62	0.88	0.85			
Medium aerosol	$1.5+5\times10^{-3}i$	0.32	0.60	0.57	} Effect depends on albedo of underlying surface	} Cooling	} Cooling
	$1.5+10^{-2}i$	0.20	0.45	0.37			
Hematite	$1.5+5\times10^{-2}i$	0.08	0.14	0.13			
	$1.5+0.1i$	0.05	0.08	0.05			
Soot	$1.5+0.5i$	0.04	0.04	0.02	} Warming	} Warming	} Warming
	$1.5+0.7i$	0.06	0.04	0.02			

[1] β—exponent of power in the size distribution of aerosol particles as given by Young: $N \sim r^{-\beta}$

[2] As the real part of the complex index of refraction for all aerosol particles has been taken as constant and equal to 1.5, identification of the particles with specific substances (including water) has been approximately estimated.

the global albedo is 0.306, which satisfactorily agrees with the data of satellite measurements (0.30). It may also be concluded that the absence of stratospheric aerosol in the case of cloud cover under it causes a warming, if the water-vapor content in the stratosphere happens to be high.

Calculations carried out by Wang and Domoto [95] showed that the effect of aerosol on thermal radiative transfer is significant. Thus for example the long-wave budget of the underlying surface decreases by 2–6% in the case of appearance of a light haze and by 7.1% in the case of a dense haze. Computations of vertical profiles of temperature, carried out on the basis of application of a modified model of radiative and con-vective equilibrium developed by Manabe and Wetherald led to the conclusion that in the presence of aerosol the intensity of convection necessary for maintaining the equilibrium considerably attenuates (by 17%) for a light and (by 28%) for a dense haze. Aerosol, however, stabi-lizes the atmosphere at the cost of an increase in the absorption of solar radiation, and also as a consequence of an increase in the mean free path of photons due to intensified multiple scattering. In the presence of a dense haze near the underlying surface, a layer of temperature inversion emerges, in which the convective transfer of heat is zero and therefore, radiation becomes a predominant factor determining heat transfer.

Atwater [22], while using the two-stream approximation, proposed a numerical model of the impact of aerosol on the albedo of the land sur-face-atmosphere system with the consideration of a variability in micro-204 structure and a complex index of refraction of aerosol. The microstructure of aerosol has been approximated by a modified gamma distribution proposed by Deirmendjian which characterizes four types of aerosols: stratospheric, oceanic, continental and urban. The optical properties of aerosol have been calculated by the Mie formula.

Atwater obtained for the above types of aerosols, relationship be-tween 'balanced' albedo (keeping in view the case, when the albedo of the system does not change with an addition of aerosol) with an imagin-ary part of the complex index of refraction (index of absorption) and the index of refraction. An analysis of these relationships showed that the selection of an aerosol microstructure considerably affects the value of the 'balanced' albedo. Thus, for example, in transition from urban to continental and from stratospheric to oceanic aerosol (for an unchanged index of refraction), the albedo decreases and a change from warming to cooling due to aerosol may occur, if the albedo of the underlying surface is close to the 'balanced' albedo.

A similar situation can be caused by variations of a real or imaginary part of the complex index of refraction. The 'balanced' albedo of aerosol satisfies the inequality:

194

$$\frac{1}{2\,[a/r+1]} \leqslant (A_s)_0 \leqslant \frac{1}{a/r+1} \, ,$$

where a and r are the absorption and reflection coefficients of aerosol, respectively. Table III.2 characterizes the discrepancies in the estimates obtained by various authors caused by a difference in the initial parameters and not by the differences in the models used.

The application of a simplified two-system approximation for the estimation of the impact of aerosol on the global albedo resulted in the necessity of investigating the correctness of such an approximation because it does not take into account the effect of the angle of incidence of solar radiation on the reflection coefficient of the atmosphere. Since the optical thickness of a cloudless atmosphere does not exceed a few tenths, Coakley and Chylek [36] proposed a method of two-stream approximation for an optically 'thin' atmosphere, in which the effect of the zenith distance of the sun has been taken into account.

Calculations carried out by this method indicate a significant effect of the scattering coefficient and zenith distance of the sun on the variability of the albedo of the earth's surface-atmosphere system. Thus, for example, large zenith angles of the sun promote a 'cooling' (increase in albedo of the system) whereas small zenith angles favor 'warming'. Hence, in 205 polar regions aerosol can show a tendency toward cooling (if we ignore the effect of molecular absorption), in spite of the higher albedo of the underlying surface.

For a correct interpretation of the warming effect, it is also necessary to take into account not only the proportion of cross sections of backscattering and absorption but also the coefficient of scattering. Comparison of results from approximate computations with data of accurate calculations showed that for an optical thickness of 0.1–0.2, the two-stream approximation is sufficiently satisfactory, even for moderately large angles of incidence of solar radiation. It correctly interprets the relationship between the global albedo and the chief parameters: optical thickness of the atmosphere, albedo of the underlying surface, zenith distance of the sun, and coefficients of scattering.

Therefore, it may be considered that the two-stream approximation is acceptable for investigations of sensitivity of climate to variations of the different parameters. Inadequacy of the considered approximation for large zenith angles would mean its unfitness for a description of scattering in clouds. The solution of such a problem is possible on the basis of application of other modifications of two-stream approximation.

In spite of serious attention being paid to the problem of solar radia-206 tive transfer in an atmosphere with aerosol, many questions remain insufficiently studied. These include the impact of horizontal and vertical

Table III.2. Estimates of 'balanced' albedo and climatic effects of aerosol

Authors	Initial values of parameters				$(A_s)_0$	Climatic effect
	$\langle \cos \vartheta \rangle$	ω	a/r	a		
Charlson and Pilat (1969)			1–4	0.1–0.3	0.1–0.5	Any
Braison and Vendland (1970)			0	1	1	Cooling
Rasul and Schneider (1971)	0.64	0.90	0.99	0.10	0.34–0.72	—do—
Schneider (1971)			2	0.08–0.15	0.17–0.33	—do—
Neumann and Cohen (1972)			0–8	0.15	0.08–1.0	Any
Sellers (1973)	0.64			0.08–0.17 0.6 (snow)	0.47	Cooling

Remarks: Here $\langle \cos \vartheta \rangle$ is a parameter of the asymmetry of the index of scattering, and ω the albedo of single scattering.

inhomogeneity of aerosol distribution on radiative transfer, the nonisothermal nature of the atmosphere, interaction between aerosol and molecular absorption and scattering, and details of the spectral structure of radiation fluxes. With the aim of assessing these effects Liou and Sasamori [64] worked out a scheme of calculation of shortwave radiation fluxes for the entire spectral band of the sun. This is based on the application of the method of discrete ordinates applied to the conditions of vertical inhomogeneity of the atmosphere and takes into account absorption by water vapor in the scattering medium (for this purpose, for every band of water vapor the equivalent coefficients of absorption have been obtained).

Specific calculations have been made for two aerosol models corresponding to conditions of clear and hazy atmospheres for horizontal ranges of visibility of 5 and 23 km and invariable vertical profiles of water vapor. In this it was assumed that the index of refraction of aerosol is not dependent on wavelength and equals 1.5, and the imaginary part of the complex index of refraction (index of absorption) linearly increases from 0 to 0.1 with an increase in wavelength in the range of 0.6–2.0 μm. Particles of aerosol were taken as spherical and homogenous, generally speaking, a nonspherical form and inhomogeneity of particles may cause a decrease in the fraction of backscattered light.

Analysis of results of computations carried out for a ten-level atmosphere shows that radiant heating of the atmosphere caused by absorption of solar radiation closely depends on the albedo of the underlying surface and the zenith distance of the sun. In the conditions of a hazy atmosphere with the sun at the zenith, radiant heating near the underlying surface (the albedo of which varies in the limits of 0.1–0.8) may attain 5–9 °C/day. With an increase in the zenith distance of the sun, radiant heating in the lower layers of the atmosphere significantly decreases.

Calculations of the albedo of the earth's surface-atmosphere system lead to the conclusion that, for an albedo of the underlying surface of about 0.3–0.4 (depending on the altitude of the sun), there occurs a transition from increase of albedo of the system with increase of aerosol content in the boundary layer (if the albedo of the surface is low) to a decrease in albedo of the system (if the albedo of the surface is high). In this connection, the optical properties of aerosol play an important role (primarily, the complex index of refraction).

An increase in aerosol content is always accompanied by an increase of aerosol absorption by the atmosphere, resulting in a warming of the atmosphere. Simultaneously, there is, however, a decrease in the total radiation influx at the level of the earth's surface (this decrease is most significant in the spectral band of 0.5–0.7 μm), i.e. a cooling of the surface takes place. The total effect is determined by the albedo of the earth's surface.

For the sun at the zenith and an albedo of the surface of 0.1 (conditions typical for the tropics), the total shortwave radiation absorbed by clear and hazy atmospheres is 27.1 and 30.6% respectively, and the fraction of radiation transmitted by the atmosphere 62.5 and 55.5%. Increase in radiation absorbed in the case of a hazy atmosphere has been caused, mainly by the absorption of radiation by aerosol in the visible and near the infrared regions of the spectrum. The albedo of the system in the considered case increases by 3.5%. If the zenith distance of the sun is 60°, and the albedo of the surface is 0.8, then in the transition from a clear to hazy atmosphere there is a decrease of 5% in the albedo of the system. Further investigations of the aerosol impact require more adequate information on the complex index of refraction of aerosol for the entire spectral band of shortwave radiation.

Analysis of the results of computation, carried out by Herman and Browning [51] shows that for a low albedo of the underlying surface, the albedo of the system may decrease or increase with increasing aerosol content depending on the complex index of refraction of particles. For a higher albedo of the surface (above 0.4), even an increase in the content of poorly absorbing aerosol would produce a decrease in the albedo of the system, i.e. brings about a warming.

A decrease in albedo always takes place, if the albedo of the surface is 0.1 or more and the imaginary part of the complex index of refraction is 0.025 or higher. In fact, the growth of the albedo of the system (cooling) is observed only when the albedo of the underlying surface is 0.1 and a complex index of refraction of 0.01 or less. If the global albedo of the surface is taken as 0.1, then the role of aerosol absorption becomes highly significant: for an imaginary part of the complex index of refraction, 0.01, there is an increase, and for a value of 0.025 there is a decrease in the albedo of the system. Since both these values lie within the limits of actual variability of the imaginary part of the complex index of refraction, this demonstrates the importance of obtaining reliable data.

If, however, the effects of cloudiness, snow and desert are taken into account, the albedo of the surface equal to 0.1 should be considered inadequate as a global characteristic. In such a case, on an average, a warming effect under the influence of an increased aerosol content should be expected for the whole earth. Naturally, in the case of nonabsorbent aerosol, the albedo of the system increases with an increase in the zenith distance of the sun. On the other hand, the converse is true for absorbing aerosol and a high albedo of the underlying surface.

The variation in optical thickness of the atmosphere exerts a determination effect on the albedo of system, whereas peculiarities of the vertical profile of coefficient of aerosol attenuation are of secondary importance. Therefore, the effect of a contamination of the boundary layer on the albedo of the system is expressed only through the amount of change in the total aerosol optical thickness of the atmosphere occurring as a result of contamination.

Raschke [78] and Kerschgens [58] computed the absorption of short-wave radiation by the atmosphere-ocean system for a model of the atmosphere. The properties of molecular and aerosol scattering of the atmosphere were in accordance with Elterman's description. In [78] calculations for two values of the index of refraction of aerosol particles, 1.5 and $1.5-0.02i$ were carried out. The latter corresponds to the albedo of single scattering, having a value of 0.84 in the entire spectral window in the $0.255-3.58$ μm range that was examined. Absorption by ozone, carbon dioxide and water vapor has been taken into account (the relative humidity in the troposphere constitutes 70%). It is assumed that the ocean is bounded by an absolutely black surface at an optical depth of the 10th interval. The entire spectral band is divided into 25 optical intervals. The extraterrestrial spectral distribution of solar radiation has been taken according to Labs and Neckel.

Calculation of the total albedo of the system for a clear sky carried out in [78] shows that, in the case of absorbent aerosol, with an increasing zenith angle of the sun, the albedo increases from 7 to 35%, 1.1 times less than in the case of non-absorbent aerosol. The presence of absorbent aerosol causes an increase in radiant heating at all levels up to 10 mb, and at the ocean surface this increase becomes 60%. The calculation of albedo of the ocean revealed a comparatively small contribution of radiation, backscattered by the layer of water, to the albedo. The results of computations for a model of the atmosphere with stratus clouds having an optical thickness of 4.0 at altitudes of 1.5–2.5 km indicate a nearly threefold increase in the albedo of the system when the sun is at the zenith, accompanied by a significant redistribution of absorption between parts of the atmosphere above (strongly absorptive) and below the clouds.

1.3. Effect of aerosol on transfer of long-wave radiation. Some of the data regarding the impact of aerosol on the transfer of long-wave radiation has already been examined in the previous section. Let us now 209 discuss this question more thoroughly, which is important, in particular, for investigating the nature of the greenhouse effect of the atmosphere.

The question of the role of aerosol as an attenuating factor for radiation in the spectral window of 8–13 μm is an aspect of primary importance concerning the problem of effect of aerosol on the transfer of infrared radiation. In this connection Pueschel and Kuhn [77] undertook investigations of the microstructure and chemical composition of aerosol in the city of Phoenix (Arizona) as well as in its rural areas with a subsequent calculation of the optical parameters of aerosol. The samples collected on membranous filter were analyzed for microstructure with the help of a scanning electron microscope. The composition of inorganic components of aerosol particles was determined by x-ray analysis. It appeared that the chemical composition of suburban samples almost coincided with the composition of the soil, and for the urban aerosol a higher content of such elements as magnesium, lead, sulfur, chlorine and iron was more characteristic. In both cases the index of refraction was taken as 1.6.

In [77], calculations of the coefficient of absorption have been carried out on the 10 μm wavelength and with the assumption that the imaginary part of the complex index of refraction (index of absorption) varies in the limits of 0.0–0.5. Along with measurements of the microstructure and composition of aerosol, helicopter measurements of vertical profiles of upward thermal radiation in the 9.5–11.5 μm band within the limits of the aerosol layer were carried out with the help of a radiometer with a 2° angle of vision. This enabled us to determine the value of radiative transmittance by the layer and the coefficient of extinction (the thickness of the aerosol layer was about 1000 m).

Since calculations showed that more than 95% of radiation attenuation in the 10 μm wavelength was caused by absorption, it appeared possible to compare the coefficient of extinction and computed values of the coefficient of absorption corresponding to various indexes of absorption. Such a comparison led to the conclusion that the index of absorption for urban and rural aerosol was 0.47 ± 0.03 and 0.19 ± 0.05, respectively, and that the volume coefficients of absorption are equal to $(34.98 \pm 0.05) \times 10^6$ and $(8.63 \pm 0.05) \times 10^7$ cm^{-1}. The data for rural aerosol agree with the results obtained earlier by Volts for aerosols and sediments soluble in water.

A major portion of aerosol consists of particles with a radius of 0.1–10 μm which significantly influence the transfer of shortwave and long-wave radiations. Therefore, Fischer and Grassl [41] compared the characteristics of absorption by aerosol in the region of the spectral window of 8–13 μm, determined for aerosol suspended in the atmosphere (on the basis of spectral measurements of the backscattering of the atmosphere) and samples of aerosol, obtained with the help of an impactor (in the second case, the index of absorption has been determined from data on the transmission by films on which the particles were deposited).

Interpretation of the data of 78 series of measurements of backscattering at four wavelengths in the 8–13 μm band, taken in Mainz (Federal Republic of Germany) from March till November, 1972, demonstrates the invariability of values of aerosol absorption observed during the said period. The intake of aerosol samples with the help of impactors with a silver chloride single-crystal base was completed during 1973–74 in Mainz and during May 1973 in the Negev Desert (Israel). In samples that were tested, particles with a radius of 1 μm and less were found to predominate.

According to measurements from Mainz the discrepancies in the values of the bulk index of refraction of aerosol k/ρ (k—index of refraction, ρ—density of aerosol particles taken as equal to 2.5 g/cm³) determined by the two methods do not exceed the errors of measurement. The specific feature of k/ρ varying in the limits of 0.05–0.15 cm³/g is the occurrence of a well-defined maximum near 9.2 μm, which is due to sulfate or quartz particles present in all types of continental aerosols. The measurements of the indexes of absorption for impactor samples, in the range of 3–17 μm, obtained at Mainz over several days in January, 1974, indicate a high variability of absorption, especially in the region of 9.2 μm.

The presence of a belt of intense aerosol absorption in the spectral window of the atmosphere significantly influences the radiant heat influx due to long-wave radiation. For the atmosphere, characterized by an optical thickness of 0.2 at the wavelength of 0.55 μm, the contribution of aerosol in radiational cooling increases from 2% in the tropics to 20% in the arctic regions. For a higher relative humidity, the contributions of aerosol and water vapor may be comparable.

The presence of aerosol leads to an increase in absorption of shortwave radiation, but the relationship with humidity in this is an inverse one (an increase in relative humidity leads to a decrease in aerosol absorption). Since heating is caused by absorption of shortwave radiation in a dry atmosphere, and predominates in a humid atmosphere radiational cooling, there should be a complete balance between heating and cooling at a certain value of relative humidity.

Interesting calculations of the spectral distribution of the intensity and fluxes of thermal radiation, taking into account the effect of mist (absorption or scattering) and water vapor (pure absorption) were carried out by Zdunkovskii and Weichel [99]. These calculations were effected using the following three models of the atmosphere: 1) exponential drop in concentration of haze particles, 2) layer of haze under inversion below the 500 m level, and 3) layer of haze below the tropopause with a thickness of 1.5 km. In all cases, the emissivity of the underlying surface has been taken as one. Calculations of the angular distribution of the intensity of thermal radiation have been carried out only for the atmospheric spectral window of 8.25–13.0 μm. Radiation fluxes and consequent variations of temperature have been computed for 3.76–92.5 μm.

Consideration of the effect of haze in the spectral window increases the intensity of downward radiation at the level of the earth's surface by

2–3 times (model 1), but has little effect on the amount of effective radiation. A dense layer of inversion haze close to the ground causes an increase in the intensity of backscattering at the zenith by 25–30 times, and leads to radical changes in the values of radiant heat exchange in this layer: in the presence of haze there is a considerable increase in radiational cooling up to 24 °C/day at the level of the upper boundary of the inversion. A significant change in the radiant heat exchange (as compared with its value in the conditions of a clear atmosphere) also takes place in the haze layer below the tropopause. Such a layer causes a significant attenuation of the upward radiation flux.

Ackerman et al. [20] carried out calculations of long-wave radiative transfer in an aerosol atmosphere. For approximate calculations of infrared radiation in a scattering and absorbent atmosphere, they proposed a four-stream approximation, based on a method of discrete ordinates which takes into account the vertical inhomogeneity of the atmosphere (ten-level model of the atmosphere in the 0–25 km altitude range). The estimates showed that the accuracy of this approximation for the calculation of flux transmission and reflection is not less than 5%.

Specific calculations have been carried out by Ackerman et al. [20] for the relationship between the complex index of refraction for aerosol particles and wavelengths obtained by Volts, when the bimodal microstructure is adopted on the basis of measurement data obtained at Los Angeles and St. Louis. The coarse fraction (modal value of radius 6–8 μm) is primarily made up of silicate particles of natural origin, and the fine fraction (modal value of radius 0.2 μm) is mainly anthropogenic (compounds of sulfur and carbon, lead, zinc, organic components, etc.).

In [20], calculations of vertical profiles of radiational cooling for models of low (mean calculated concentration of particles is 10^{-1} cm^{-3}),

Table III.3. Optical parameters of atmospheric aerosol

Parameter		Model of atmosphere having dust content		
		low	moderate	high
n	coarse fraction	2.63	2.63	2.63
k		0.80	0.80	0.80
n	fine fraction	1.98	1.98	1.98
k		0.11	0.11	0.11
ω_a		0.4452	0.3827	0.3713
ω_s		10^{-5}	0.1977	0.3371
$\tau(1)$		0.0788	0.227	1.395

Remarks: Here n, k are real and imaginary components of the index of refraction $m=n-ik$, respectively, ω_a, ω_s are albedo values of single scattering for aerosol and an aerosol-gas system, and $\tau(1)$ is the optical thickness of a 1 km layer of the atmosphere.

212 moderate (10^5 cm^{-3}) and heavy (10^6 cm^{-3}) dust contents in the atmosphere are given. The parameters presented in Table III.3 characterize these models.

Naturally, the impact of aerosol attenuation on radiational cooling is practically absent in the region of molecular absorption bands, but is strongly manifest in the spectral windows of the atmosphere. Table III.4 indicates the rates of radiational cooling relating to the conditions of low and moderate dust contents of the atmosphere in separate intervals Δz of the lower 3 km layer of the atmosphere, and clearly illustrates this point.

Table III.4. Dependence of radiational cooling on atmospheric dust content

Bands	Δz, km	Rate of cooling (K/day) when dusty		Variation of radiational cooling
		low	moderate	
15 μm	1.75	−0.103	−0.103	0.0
6.3 μm	1.75	−0.550	−0.625	−0.075
Rotational	1.75	−0.479	−0.526	−0.047
Window	1.75	−1.008	−1.328	−0.320
Window	1.00	−1.667	−2.494	−0.827
Window	1.00	−1.667	−5.922	−4.255
			(high dust content)	

Analysis of vertical profiles of radiational cooling indicates that, in the case of a low dust content in the atmosphere, cooling near the earth's surface dominates; in the conditions of a moderate dust content there is intense cooling over the entire boundary layer, and a sharp peak of cooling (-11.7 K/day) near the 0.5 km layer is observed for a high dust content. It may, therefore, be considered that an aerosol layer under inversion should promote inversion and thereby support a further increase of aerosol concentration. An increase in the dust content of the atmosphere causes a significant decrease in the effective radiation of the underlying surface.

213 The conclusion drawn here is confirmed by a great deal of observational data. Thus, for example, Idso [54] carried out on May 15, 1973, (in Phoenix, Arizona) measurements of the radiation budget of a bare soil surface and its components in conditions of a high transparency of the atmosphere (horizontal range of vision—67 km) and for a mild dust storm, which appeared after midday and had reduced the visibility to 11 km (dust suspension extended up to the altitude of about 2000 m and remained after sunset). Determination of the relative emissivity of the atmosphere enabled us to reveal its increase by 10% after the atmosphere

became dusty. However, a measurable decrease of total radiation could not be revealed. Hence, the radiation budget increased when the atmosphere became dusty. On these grounds it may be assumed that an increase in the anthropogenic aerosol content in the atmosphere may initially cause a warming rather than a cooling of the global climate, as is usually considered.

A vast amount of information regarding the effect of atmospheric dust on long-wave radiation transfer has been obtained from the data of ship and aircraft observations during the GATE [13, 16] expeditions. As already mentioned, in the first phase of the GATE (June 24, July 10, 1974) intense removal of dust from the Sahara by harmattan winds was observed. The boundaries of the haze layer were located near the 980 and 520 mb levels (0.6–6.25 km).

Kuhn et al. [61] measured vertical profiles of long-wave radiation in the range of 8.0–14.0 and 9.5–11.5 μm bands with the help of a radiometer with an angle of vision of 2° installed on the *Convair 990* aircraft laboratory on June 30, July 3 and 29. The aim of the measurements was to analyze the effect of haze on long-wave radiative transfer and develop an approximate method for computation of long-wave radiation fluxes and influxes in a dust-laden atmosphere. According to radiometric data for measurements of temperature of the ocean surface, the surface temperature field in the investigated belt of 8–14°N was fairly homogeneous.

According to the approximate method of calculations (in agreement with observational data) the average rate of radiational cooling for the entire dust layer is 0.09 °C/h, whereas in a similar layer of the dust-free troposphere the radiational cooling was 0.06 °C/h. The mean bulk scattering coefficient for the layer of haze, which was fairly homogeneous along the height, was 0.042 km^{-1}. The increase in radiational cooling in the dust layer is not large and may appear to be balanced by an increased absorption of shortwave radiation. Near the ocean surface, the effect of a dust layer was generally not observed. Therefore, it is important to note that a decrease in shortwave radiation influx caused by haze leads to an overall cooling. It is extremely important to compare the haze effects in the long-wave and shortwave bands on the basis of measurement data.

1.4. Aerosol and dynamics of the atmosphere. Problems related to the investigation of radiation and dynamics of the atmosphere from the standpoint of aerosol effects have attracted much attention in recent years. In connection with the solution to a similar problem, relating to the problem of the boundary layer of the atmosphere, Welch and Zdunkovski [97] proposed a model for the composition and structure of a polluted boundary layer of the atmosphere, as well as a method for calculation of radiative transfer in similar conditions for the spectral band of 0.29–100 μm, including the spectra of shortwave and long-wave radia-

204

tions. It is assumed that, within the limits of the boundary layer of the atmosphere of 3 km thickness, the shortwave radiative transfer is determined by the effect of aerosol scattering and absorption, molecular scattering, and absorption by water vapor and nitrogen dioxide (pollutant).

Calculations of radiant heat influx showed that, because of absorption of solar radiation, radiant warming is maximum near the upper boundary of radiation inversion and attains 2 to 4 K/h for low and high humidities, respectively, for a zenith angle of the sun at 45°. For larger zenith angles, radiant heating decreases. As a rule, the contribution of aerosol to radiant warming considerably exceeds that of nitrogen dioxide.

The fluxes and shortwave radiation influxes near the earth's surface closely depend on humidity. Ignoring the effect of humidity on the coef-
215 ficients of absorption and aerosol extinction gives rise to errors in the computation of the shortwave radiation budget of the underlying surface extending up to 40%. The consideration of aerosol is also important in the calculation of long-wave radiant heat influx. Neglect of the effect of humidity on the optical characteristics of aerosol causes errors in calculations of the effective radiation of up to 35%. Though the aforementioned results pertain to the conditions of winter, they should also be similar for summer. Apparently, the effect of humidity should also be taken into account in the study of the radiation effects of continental aerosol above the ocean.

Consideration of the impact of aerosol and gaseous pollution on the dynamics of the boundary layer is especially significant in the case of radiation inversion. It is well known that during periods of sharp inversion at the ground level there is an accumulation of a large amount of aerosol and gaseous pollutants in the inversion region of the boundary layer of the atmosphere. This leads to an increase of backscattering in the atmosphere (and consequently to a decrease in night cooling of the underlying surface) as well as to variations in the value of radiant heat influx in the boundary layer. As a result of a complicated interaction of these processes with a turbulent mixing and heat exchange in the soil, vertical profiles of temperature and wind may be observed even during calm nights. After sunrise, the effect of aerosol and gaseous pollutants leads to a decrease in the total radiation, which in turn slows down the effect of radiation inversion. Absorption of solar radiation by pollutants appears to be an additional factor affecting the profiles of temperature and wind.

An interesting example of anthropogenic aerosol effect in the boundary layer on air temperature was discussed by Galindo and Chávez [43]. In recent years, in Mexico it had become a common practice to burn discarded automobile tires in the streets at Christmas time and on New Year's eve. This resulted in a considerable release of aerosol and gaseous

pollutants into the atmosphere, which in the presence of a stable stratification causes a severe pollution of the atmosphere (municipal bodies
have now decided to curb such practices). In [43] the authors have
analyzed the conditions of atmospheric pollution created on December 25,
1977 during the daytime (7–14 hours of actual solar time) after the burning of automobile tires on Christmas eve. The analysis was carried out
on the basis of observational data of the Central Meteorological Observatory in Takubaja (19° 24′N, 2308 m above sea level) and measurements
of solar radiation in the university campus located on the outskirts of
Mexico city (19° 20′N, 2268 m above sea level).

216   The month of December in Mexico is characterized by cold, clear
weather with winds raising enormous amounts of natural dust in the air.
The data of aerological soundings over 6 hours often reveals a temperature inversion on the ground (this relates to the day under consideration).
The Richardson number on December 25 was 48094, which indicates an
almost complete absence of turbulence and the presence of stagnant air.
During the period from 07 : 00–14 : 00 h, the horizontal range of visibility
was observed to be 1 km (at midday, it increased to 5 km) and from 12 to
14 hours it was only 0.1 km. By 15 : 00 h visibility again increased to
6 km. The height of the layer of mixing, in the morning, was 90 m and
was accompanied with a wind speed of 1 m/s.

The presence of an aerosol layer of more than 2 km thickness caused
an increase in night temperature by 2–3 °C, and a decrease during the
daytime by 2 °C. The duration of solar illumination decreased by 1 h.
Haziness in the atmosphere attained a very high level (the Angström
coefficient of haziness rose 68% above the mean value). The total radiation decreased by 26.8% and scattering increased by 12%. The ratios of
scattered and direct solar radiation with respect to total radiation were
58 and 42% respectively. On the whole, the smog layer behaved like
clouds of the middle level in which aerosol absorption prevails over
scattering.

An interesting model of radiation processes in the layer of atmosphere,
polluted by aerosol, has been developed by Bednar [23]. In this, a system
of equations of thermal conductivity determining the variation in the
temperature of air, underlying surface and rock with the consideration
of a radiant and turbulent heat influx in the air and a conductive heat
transfer in the soil has been formulated. The influx of heat because of
shortwave and long-wave radiations has been determined by applying
the simplified variant of the two-stream approximation for the equation
of radiative transfer.

In [23] calculations relating to the effect of a ground aerosol layer of
300 m thickness on the thermal regime of this layer during day and night
have been carried out. In this it was assumed that particles of aerosol are

purely absorbent in nature. The total radiation at the level of the upper boundary of the layer has been given as 840 W/m², which included back-scattering amounting to 260 W/m². The bulk coefficient of aerosol extinction (exponentially decreasing with altitude) is 10^{-4} cm^{-1} at the level of the underlying surface. The albedo and emissivity of the surface are equal to 0.2 and 0.8 respectively. The coefficient of mixing was taken as 0.16 cm²/s, which is independent of altitude and corresponds to con-
217 ditions of an absolutely calm atmosphere (molecular viscosity), i.e. the thermal regime is determined only by radiant heat influx.

An analysis of the tables reproduced in detail in [23], interpreting the results of calculations, shows that during the daytime the aerosol layer is heated due to absorption of shortwave radiation. This intensifies the instability of the atmosphere and promotes an upward transport of aerosol pollutants. The effect of aerosol is less clear during the night: at higher concentrations of aerosol, cooling intensified (determined by the effect of radiant heat exchange), which promotes the formation of inversion above the aerosol layer and an increase in stability of the atmosphere. Further development of the model described is possible by taking into account turbulent mixing, influx of heat due to phase transformations of water in the atmosphere, and the effect of gaseous components of the atmosphere on radiative transfer.

Within the framework of a one-dimensional model of radiative and convective scattering developed by Manabe and Wetherald, Reck [79–82] studied the impact of aerosol on the temperature of the underlying surface. The two-stream approximation with the consideration of aniso-tropic scattering, proposed by Sagan and Pollack, was used for describing the effect of aerosol on radiative transfer. In such a case the two basic optical parameters of aerosol are: the albedo of single scattering and the anisotropic scattering factor. The complex index of refraction was taken as $1.5-0.1i$ (calculations showed that the thermal effects of aerosol change very little if the imaginary part of the complex index of refraction varies in the limits of 0.01–0.1).

In [82] the authors present results of computations of temperature variations ΔT_s of the earth's surface caused by the effect of aerosol depending on the coefficient of extinction σ_v in the wavelength of $\lambda = 0.55$ μm for three values of the underlying surface albedo A_s (0.07, 0.3 and 0.6). Naturally, the impact of aerosol is strongly felt only when it is located in the lower layers of the troposphere over a dark underlying surface ($A_s = 0.07$). In this case, the drop in surface temperature caused by aerosol at $\sigma_v > 3$ is approximately 7 K. The heating effect by aerosol is observed only over surfaces with a higher albedo (just as in the case of snow), except where $\sigma_v > 0.9$, when a warming takes place at $A_s < 0.6$.

Calculation of the ratio between the values of A_s and σ_v for mean global conditions under which the transition from cooling to heating (or the converse), showed that for a greater optical thickness ($\sigma_v \to \infty$), the limiting value of A_s approaches 0.35. For a range of the most typical values of optical thickness, the critical albedo is approximately equal to 0.6. It is significant that these estimates are not sensitive to the presence of water-bearing clouds.

Reck [82] analyzed the results of computations of ΔT_s, characterizing the relationship $\Delta T_s (A_s)$ for various latitudinal belts at aerosol optical thickness τ equal to 0.065 and 0.26. In the limits of 5–35°N there was a monotonic attenuation of cooling with an increase in A_s and a small latitudinal difference. At 55°N and further north, a latitudinal difference was more clearly observed, but an increase in cooling was even more significant. The increase in cooling was almost twice that in the zone of the maximum for $A_s = 0.3$; it was even more considerable at $\tau = 0.26$.

Thus an increase in aerosol content accelerates cooling as the critical value approaches during a variation from any side of the annual course of the albedo, which in its turn, influences the transition from winter to summer. During spring, melting is delayed, and in autumn freezing is accelerated, because of which the duration of winter is prolonged and the extent of ice cover increases. The probable empirical confirmation of this conclusion may come from satellite observational data, which revealed an increase in the extent of ice cover in 1971. In [82] it is emphasized that realistic estimates of aerosol effects are possible only with the availability of reliable observational data.

On the application of schemes of radiative transfer with a few parameters to the theoretical models of climate, the necessity of regular observations of corresponding parameters has emerged. In order to obtain information on the optical properties of atmospheric aerosol, regular observations were undertaken at a number of points in the USA with the help of nephelometers on the 530 nm wavelength. The observations enabled us to determine the aerosol coefficient of scattering b_{sp} and the ratio R of backscattered flux to scattering in all directions [21]. Apart from this, by using filtered samples, measurements of the coefficient of absorption of aerosol b_{ap} permitting the determination of the absorption coefficient with an accuracy up to a factor of two, were carried out. Sometimes these measurements were carried out separately for particles having a radius smaller or greater than 2 μm.

Measurements showed that absorption by small particles is 4–32 times more intense than in the case of large particles (on the average by 16 times) but the mass of the fine-dispersed fraction is greater by 2–4 times only. The values of b_{ap} for all particles vary in the limits $(0.07-0.7) \times 10^{-4}$ m^{-1}, the values of b_{sp} in the limits $(0.12-1.8) \times 10^{-4}$ m^{-1} and of R from 12 ± 1 to $20 \pm 8\%$. Both, on a global scale and in regions of the eastern USA and northern Europe, the sulfur component dominates in the composition of aerosol.

The most probable absorbent components of aerosol seem to be the oxides of metals (especially hematite and magnetite), elementary oxygen and possibly some large-molecular compounds. Since removal of metallic oxides and large-molecular compounds from the samples did not lead to noticeable decreases in the coefficients of absorption, it should be considered that the chief absorbent component is oxygen. The calculations of the most significant optical parameters of aerosol led to the conclusion that in the cases considered the albedo of single scattering ω_0 varies within the limits 0.53–0.87 and the parameter $\chi = (1-\omega_0)/\omega_0 R$, characterizing the aerosol effect of heating or cooling varies from 0.83 to 5.20. For a typical albedo of the underlying surface equal to 0.2, the critical value of χ is 2. From this it follows that for example in St. Louis and Arkansas there is aerosol heating and in Denver, on the contrary, aerosol cooling prevails.

In survey [9], estimates of aerosol effect on climate have been briefly reviewed. Without repeating these results, it can only be stated that till now the estimates of climatic effects of aerosol were based, as a rule, on the application of one-dimensional globally averaged models of radiative or radiative-convective equilibrium as well as models of local effects (effects of pollution).

Joseph [57] was the first to investigate the effect of climate within the framework of a three-dimensional model of general circulation of the atmosphere, developed at the US National Center for Atmospheric Research (NCAR). The fundamental assumption of the model of desert aerosol includes the global distribution of sources of aerosol and the Eddington delta-approximation, which describes the shortwave and long-wave radiative transfer. In order to improve the signal-to-noise ratio, the aerosol content has been raised 10 times as compared to the normal value while numerical modeling of the climate has been carried out to take into account the impact of aerosol on the transfer of 1) only short-wave radiation, and 2) both shortwave and long-wave radiation. The aerosol effects, expressed as deviations of localized and zonal averaged values of temperature, and vertical and meridional components of velocity from control values, have been considered. It may be noted that all the values have been determined as averages over the last 30 days of the experiment for the steady state relative to the conditions of July.

In [57], the global maps of these deviations showing for example, that in the first of the foregoing alternatives there exists a negative anomaly in the underlying surface temperature of a few degrees in all deserts (deserts of the USA, however, are an exception), have been reproduced. The heating of a 3–6 km layer of the atmosphere up to 11 K is observed over the Atlantic Ocean where a dense aerosol layer is present in the free atmosphere. Since aerosol heating of the atmosphere over continental areas for the same aerosol optical thickness, is of only a few degrees, it indicates the importance of consideration of the peculiarities of the vertical distribution of aerosol.

Calculations for the second alternative also indicate a temperature drop caused by aerosol, but in the given case this effect is expressed most severely in the Middle East, Central Asia and Australia and not in the Sahara. The effect of aerosol on the thermal regime of the free atmosphere, in this case, appears to be poorly expressed and this especially relates to the Atlantic Ocean. Calculations of zonal averaged values for the first alternative indicate a drop in surface temperature in deserts and a rise of air temperature above deserts as well as a significant transformation in the field of vertical movements.

Consideration of the effect of aerosol on shortwave as well as longwave radiative transfer leads to an increase in averaged zonal values of temperature especially in temperate latitudes. Transformation of the field of vertical movements is even weaker. On the whole, the impact of aerosol on the global circulation of the atmosphere, both near the source of aerosol as well as beyond its limits, should be established. Differences in the effects with two variants of aerosol under consideration demonstrate the importance of the application of adequate information on its spectral properties.

It is important to bear in mind that according to computational data for a one-dimensional model, the aerosol content has increased 10 times as compared to its normal value. The variations of temperature for a normal aerosol content are only 3 times lower than when the aerosol content is 10 times higher. This means that the effect of aerosol on the global circulation of the atmosphere should be significant even with a normal aerosol content.

Estimates of climatic variation within the framework of a semi-empirical (energy budget) model depend to a great extent, on the consideration of variations of the earth's albedo as a leading factor in climatic variations, and of albedo inversion relationships as determined by a severe contrast in the albedo of snow or ice cover and exposed water. Henderson-Sellers and Meadows [52] carried out calculations of the mean annual albedo of the earth as a whole, bearing in mind the application of the
221 results obtained for an analysis of the contribution of the variations of the following three factors to the variability of the global albedo: the albedo of the underlying surface, clouds and atmospheric aerosol.

Calculation of the mean annual global albedo of the earth is carried out under the assumption that the axis of rotation of the earth is perpendicular to the direction of incident radiation. As a standard model the global field of cloudiness concentrated (in accordance with climatic data) in the 40–77°N and 0–19°S zones for 50% of the clouds symmetrically distributed in equatorial and temperate latitudes of both the hemispheres has been given. The polar ice caps can extend to ± 60° latitudes.

A global albedo of 30.2%, for albedo values in the northern and southern hemispheres equal to 29.6 and 30.7%, respectively, corresponds to the model under consideration. Insofar as it is usually considered that the most severe impact on albedo is exerted by the polar caps, the earth's albedo was calculated under the assumption that the polar caps shift latitudinally by ± 20°. Such variations however, led to negligible variations in the albedo of the hemispheres and the earth as a whole (Table III.5). A more significant change in the albedo takes place if we assume that the ice cover for an unchanged extent is covered with snow.

Table III.5. Global albedo and its variations (as compared with 'standard' earth) with variation of ice cover and clouds (φ—latitude)

Conditions of observations	Albedo, %			
	global	northern hemisphere	southern hemisphere	variation
'Standard' earth	30.2	29.6	30.7	0
Ice cover				
$\varphi > 80°$	30.0	29.5	30.6	−0 7
$\varphi > 40°$	30.2	29.6	30.7	0.0
Snow cover, $\varphi > 60°$	30.5	30.0	31.0	+1.0
Belts of cloudiness at temperate latitudes symmetrically displaced by				
5° toward poles	29.6	29.0	30.1	−2.0
5° toward equator	31.1	30.6	31.6	+3.0
Global cloudiness				
60%	34.9	34.5	35.3	+15.6
50%	30.2	30.1	30.2	0.0

222   A deterministic effect on the albedo of the earth is exerted by variations of cloudiness: increase in the amount of clouds up to 60% or even a slight displacement (by 5° latitude) of the belt of cloudiness in the temperate latitudes toward the poles or the equator (Table III.5). Since observations demonstrate a near equality of albedo of the hemispheres, calculations have been carried out for the conditions of a slight asymmetry in the amount of clouds in the hemispheres, which showed that such an asymmetry is enough for equalizing the albedo of the hemispheres. The changes considered for zonal distribution of clouds have taken place in the geologic past and, consequently, are not excluded in the future.

This shows the importance of the inverse relationship caused by variations in the amount of clouds in numerical modeling of the climate. Approximate estimates of the effects of variations of dust content in the atmosphere, for conditions after eruption of Mt. Krakatau in 1883, on the albedo in the northern hemisphere could not have exceeded 0.5%.

The progress in numerical modeling of the climate requires a reliable account of the effect of aerosol based on the development of models of aerosol and assessment of the sensitivity of climate to various characteristics of aerosol. In view of this, the following aspects of the problem must be primarily explained [20a]: 1) effect of aerosol on the processes on global and regional scales, 2) identification of the most significant (from the viewpoint of impact on climate) types and properties of aerosol, and 3) comparison of effects due to variability in albedo and minor gaseous components (water vapor, oxygen, carbon dioxide etc.), cloudiness and albedo of the underlying surface.

In an attempt to systematize the properties of aerosols it is worthwhile to classify these in six categories: 1) mineral dust blown by winds, 2) products of evaporation of oceanic jets, 3) anthropogenic (industrial) aerosol, determined by the ejection of particles and transformation of vapor into particles, 4) volcanic aerosol (ejection of particles and transformation of gas into particles), 5) haze formed during combustion of biotic components on the continental areas, and 6) stratospheric aerosol.

The most comprehensive characteristic of aerosol properties can be obtained by a systematic determination of its concentration, microstructure, chemical composition (depending on the size of particles) and index of refraction, and also the optical parameters: optical thickness, albedo of single scattering and indexes of scattering in relation to wavelength. In this it is important to take into account the fact that, as a rule, the microstructure of aerosol in bimodal and each mode characterizes specific optical properties.

In order to take into account the effects of aerosol in the theory of climate, the development of models of dynamics of formation, transformation and transfer of aerosol in the interaction with the processes of climate formation, is of great importance. The combined application of data from ground, aircraft, balloon and satellite observations should constitute a basis for aerosol climatology (models), while conducting the complex program of investigations with the use of an entire set of devices for observations has acquired a special importance.

Numerical experimentation on the sensitivity of climate to the aerosol effect should be conducted keeping in view the sequential revelation of the impact of aerosol on the radiation budget of the earth's surface-at-mosphere system, influx of heat, thermal budget and climatic system as a whole. A study of the interaction between dust aerosol and clouds, is important. At the same time, the effect of mineral dust aerosol on the transfer of long-wave radiation requires serious attention.

In order to obtain approximate evaluation of the climatic effect of aerosol, models of radiative-convective equilibrium are suitable. With the help of these models, it has been possible to establish the marked effect on the vertical profiles of temperature for a high content of aerosol and a very weak influence of aerosol when the optical thickness is less than 0.5. Depending on the albedo of the underlying surface, the stratification of aerosol and clouds can give rise to cooling as well as warming effects (up to a few degrees).

Stratospheric aerosol can exert a considerable influence on climate. The consideration of the aerosol effect in the framework of a zonal model of general circulation of the atmosphere led to the conclusion regarding a drop of surface temperature by a few degrees. The three-dimensional modeling of general circulation of the atmosphere while taking desert aerosol into account reveals the following: 1) a drop in surface temperature in deserts up to 2.5 K if it is assumed that aerosol absorbs only solar radiation, 2) an increase in surface temperature in the desert up to 3.5 K if the effect of aerosol on the transfer of thermal radiation (greenhouse effect) is taken into account, 3) a considerably higher stability of the atmosphere near the ground and attenuation in the free atmosphere (in the upper half of dust-laden zones of the atmosphere), and 4) a significant transformation of the field of temperature beyond the limits of dust-laden zones of the atmosphere (Western Europe, Asia and tropical Africa).

2. STRATOSPHERIC AEROSOL

224 The problem of stratospheric aerosol merits special attention in view of its long residence time, which in particular, has given rise to the question about the possibility of influencing climate by manipulating the quantity of stratospheric aerosol [2].

In order to find a solution of the problem concerning the effect of stratospheric aerosol on climate, the investigations of processes leading to the formation of aerosol in the stratosphere are of prime importance. The works of Turko et al. [93] and Toon et al. [94] have made a significant contribution to these researches.

The presumed effect of the stratospheric sulfate-bearing aerosol layer on the climate, discovered by Young et al. in 1961, and the possibility of an anthropogenic influence on this layer have enhanced interest in investigations on aerosol in the stratosphere. One of the important objectives of such investigations is the development of methods for estimating the effect on stratospheric aerosol. Bearing in mind the search for a solution to this problem, Turko et al. [93] developed a one-dimensional nonsteady

model describing the evolution and formation of aerosol in the stratosphere. The model fairly well simulates the observed properties of the
stratospheric aerosol layer.

The results of numerous direct and remote-controlled measurements
enable us to present a relatively complete picture of the aerosol layer.
The maximums of number and mass concentration of particles with a
diameter of more than 0.15 μm usually occur at altitudes of 20–25 km,
being closely linked with the altitude of the tropopause. The stratospheric
aerosol layer is characterized by the presence of annual and latitudinal
trends, as well as a short-term variability, and it is considerably strengthened after volcanic eruptions. For a microstructure of aerosol, a predominance of minute particles is typical: the ratio of concentrations of
particles with diameters greater than 0.15 and 0.25 μm lies in the range
of 3–5. However, the concentration of particles with radii less than 0.1
μm rapidly decreases. Differing from the concentration of particles with
a diameter greater than 0.15 μm, the total concentration of all the particles rapidly decreases with increasing altitude above the tropopause.

The earlier theoretical investigations showed that stratospheric aerosol can emerge from gaseous components of sulfur, whereas the sulfates
formed as a result of a corresponding reaction are sufficient for maintaining the observed stratospheric aerosol layer. In [93] a nonsteady one-
dimensional model has been proposed with the purpose of obtaining a
more comprehensive understanding of aerosol physics and photochemis-
225 try of sulfur than were hitherto examined in various investigations. This
has been done by the application of the modern theory of nucleation
and growth of aerosol particles, as well as the latest data on the content
and reactivity of sulfur compounds in the air.

The presence of solid nuclei identified from observational data (about
30% of the particles contain such nuclei), which is of significant importance for the description of accumulation and evolution of stratospheric
drops of a solution of sulfuric acid, has been kept in mind. Analysis of
the size and chemical composition of nuclei can be an important tool for
understanding the behavior of the stratospheric aerosol layer.

Numerical modeling covers the 30-layer thickness of the atmosphere
(0–58) km taking into account the tropopause as a source of gases and
nuclei for condensation as well as a sink for aerosol droplets. However,
the physics and chemistry of tropospheric aerosol have only been studied
in a rough approximation. The microstructure of aerosol is assumed
through a distribution function of particles of different radii within the
limits of 0.01–2.56 μm. This is done by classifying the sizes into 25 grades
in such a way that during transition from one into successive grades the
volume of particles doubles.

In the process of diffusion, the gaseous compounds of sulfur in the stratosphere undergo oxidation as a result of chemical and photochemical reactions. This combination of reactions results in the formation of sulfuric acid vapors. Observations confirm that when there exist many gaseous components containing sulfur in the lower troposphere, in the upper troposphere only sulfur dioxide and COS are found in noticeable concentrations. Therefore, only these components are considered as sources of sulfur for aerosol particles.

Regarding sulfur dioxide, it is accepted that its concentration on the earth's surface is 0.5 ppbv and reduces to 0.05 ppbv of carbonyl sulfide at the level of the tropopause due to leaching by precipitates. The chemically stable carbonyl sulfide (apparently its residence time in the tropopause is several years or more), the concentration of which in the troposphere has been assumed to be 0.47 ppbv, penetrates into the stratosphere in the form of cholorofluorocarbons, where it is subjected to photolysis $(COS + h\nu \rightarrow S + CO)$ with subsequent formation of sulfur dioxide:

$$S + O_2 \rightarrow SO + O;$$
$$SO + O_2 \rightarrow SO_2 + O;$$
$$SO + O_3 \rightarrow SO_2 + O_2;$$
$$SO + NO_2 \rightarrow SO_2 + NO.$$

226   Calculations have provided values of the global flux of sulfur dioxide through the tropopause at 2×10^5 t/yr. The disintegration of COS is equivalent to 9×10^4 t of sulfur dioxide in a year. The chief reactions of sulfur dioxide oxidation leading to sulfuric acid are:

$$SO_2 + O + M \rightarrow SO_3 + M;$$
$$SO_2 + H_2O \rightarrow SO_3 + OH,$$
$$SO_2 + OH + M \rightarrow HSO_3 + M;$$
$$HSO_3 + H \rightarrow SO_3 + H_2O;$$
$$SO_3 + H_2O \rightarrow H_2SO_4.$$

Insofar as the rates of the first two reactions are low (time constants are of the order of months or years), a faster third reaction dominates in the lower stratosphere. Hence, in the model considered, the following compounds of sulfur have been discussed: S, SO, SO_2, SO_3, HSO_3, H_2SO_4 and COS. In this the vertical profiles of concentration of oxygen, ozone, hydroxyl, water vapor and nitrogen dioxide are given. At certain altitudes, the supersaturation of the H_2SO_4–H_2O gaseous mixture takes place with a deposition of a solution of sulfuric acid vapor on the surfaces of condensation of the nuclei. The drops of solution thus formed, being situated within the medium and being supersaturated relative to sulfuric acid vapors, grow due to molecular heterogeneous condensation of water and acid vapors. Freshly formed particles apparently consist of nuclei surrounded by an aqueous solution of sulfuric acid (possibly it is a pure solution) with a concentration (by mass) equal to 70–80%.

It is assumed that the nucleation of the Aitken nuclei, being of tropospheric origin and located in the layer of supersaturated vapor of sulfuric acid, takes place in 10^6 s. It is also accepted that the concentration of nuclei on the earth's surface is 1200 cm^{-3}, and their microstructure is described by an r^{-4} proportionality with the minimum radius of particles r of a value of 0.01 μm. The growth of drops, caused by heteromolecular condensation of vapors of sulfuric acid and sulfur dioxide which was studied earlier by Hamil et al., is determined by the rate of entry of molecules of sulfuric acid into the drop. The drops also undergo a coagulation due to Brownian movement, precipitate under gravitational force and are diffused in a vertical direction. Below the tropopause, a leaching of particles by precipitates occurs.

Computation of the amount of solid (or dissolved) nuclear matter in the particles of aerosol, on the basis of models, enables us to reveal the conditions of growth or evaporation of drops depending upon local environmental conditions. When the drops evaporate, their nuclei remain 227 in the form of solid particles. A kinetic equation thus obtained, makes it possible to describe the space-time fluctuations of concentration of aerosol droplets and condensation nuclei, as well as dimensions of nuclei in the droplets with the consideration of change in the concentration of aerosol particles due to nucleation, condensation growth (or evaporation), coagulation, sedimentation, diffusion and leaching.

Numerical modeling, developed by researchers, is applicable to the conditions of steady stratospheric aerosol without any element of manipulation by way of varying individual parameters. A comparison between results of computational and observational data revealed a fairly satisfactory correspondence—a fact that demonstrates the possibility of the application of a one-dimensional model for the description of general characteristics of the aerosol layer. In particular, the calculated and observed values of mass concentrations of sulfate-bearing aerosol, vertical profiles of particle concentration (with particles of radii greater than 0.15 μm), and data on the microstructure of aerosol, show good agreement.

For a correct comparison with experiment, data of comprehensive observations which still does not exist, is required. The model of stratospheric aerosol is based on various assumptions that have been discussed in detail. The most serious limitation of the model is its one-dimensional structure and the consequent disregard for the horizontal transport of gases and particles. In view of this the development of a two-dimensional model based on the application of existing two-dimensional models of minor gaseous components of the stratosphere is being planned. The use of a one-dimensional model has been proposed for the study of the role of volcanic eruptions and other sources of sulfur in maintaining the stratospheric aerosol layer, as well as for the estimation of the contribution of anthropogenic excretions, releasing sulfur-containing gases as well as solid particles.

Calculations have been carried out with the aim of analyzing the sensitivity of the theoretical model of formation and evolution of stratospheric aerosol to the selection of parameters and to reveal the most significant factors characterizing the effect of nucleation, growth, evaporation, coagulation, diffusion and sedimentation. The objective of numerical modeling under consideration is also to justify the program of experimental and theoretical investigations essential for a broader study of the processes governing the formation and evolution of stratospheric aerosol.

The general characteristics of studied factors and their influence on the various properties of aerosol are presented in Table III.6. These properties include the total calculated ratio of mixture of particles with radii greater than 0.15 μm, the bulk ratio of the mixture of sulfur-bearing aerosol and the ratio of concentrations of particles with radii greater than 0.15 and 0.30 μm.

229   Table III.6 demonstrates that the total ratio of the mixture is determined mainly by the effect of coagulation, but at the same time the concentration of condensation nuclei at the level of the tropopause and the coefficient of diffusion at higher altitudes also have some importance. The bulk ratio of the mixture and the ratio of mixture of coarse particles is regulated by the processes of growth, sedimentation and also by the

228   **Table III.6.** Sensitivity of various characteristics of stratospheric aerosol to factors determining these characteristics

Factors	Total calculated proportion of mixture	Proportion of mixture of coarse particles	Mass proportion of mixture of sulfur-bearing aerosol	Proportion of concentration
I. Sources and sinks of gases:				
Variable concentration	a	h	h	h
Release of SO_2 in stratosphere	a	h	h	m
Concentration of HSO_3	a	a	a	a
Variable concentration of OH	a	a	a	a
II. Time for deposition of particles and influx of nuclei:				
Change in the altitude of tropopause	m	h	h	h
Variation of coefficient of diffusion	m	m	m	a

Change in the time of leaching in troposphere	a	a	m	m
Increase in sulfuric acid drop in troposphere	a	m	a	a
Variation of condensation nuclei concentration at level of tropopause	m	m	a	m
Variation of dimensions of condensation nuclei at level of tropopause	a	a	a	a
Termination of sedimentation	a	h	h	m
III. Nucleation:				
Variation of time of heterogeneous nucleation	a	a	a	a
IV. Growth:				
Variation of stratospheric temperature	a	h	m	m
Variation of H_2O concentration in stratosphere	a	a	a	a
Variation of vapor pressure of H_2SO_4	a	m	m	m
Variation of coefficient of absorption of H_2SO_4 molecules	a	a	a	a
Termination of hetero-molecular growth	a	h	h	h
Termination of coagulation	h	a	a	a
Termination of rise in H_2O concentration	a	a	a	a

Remarks: h—high sensitivity (sensitivity of considered characteristics of aerosol to the studied factor exceeds observed variability), m—moderate sensitivity (within the limits of observed variability), a—absence of sensitivity (calculated variability less than observed).

removal of particles on account of evaporation at high altitudes and leaching in the troposphere. Such factors as the strength of the sulfur dioxide source and time of deposition of aerosol in the stratosphere exert a more significant effect on the bulk ratio of the mixture of coarse particles than the influx of condensation nuclei. If the microstructure of particles with a radius of 0.1–1 μm is determined, particularly, by the time of deposition and intensity of the sulfur-dioxide source, in that case coagulation becomes a dominant factor for extremely fine particles, whereas concentration of nuclear condensates is the deciding factor in the case of coarse particles.

The observed values of SO_2 and COS concentrations in the troposphere are sufficient for maintaining the undisturbed layer of stratospheric aerosol, if their transfer through the tropopause alone is taken into account. The contribution of carbonyl sulfide is decisive for an unperturbed layer and the contribution of sulfur dioxide may be significant for direct deliveries into the stratosphere. Sulfur dioxide exerts a significant effect on the number of particles directly above the tropopause where the calculated (volumetric) concentration and not the ratio of mixture happens to be the highest. The main perspective of furthur development of the theoretical model of stratospheric aerosol is related with the setting up of the two-dimensional model (it provides an opportunity to study latitudinal variations), consideration of interaction between acid droplets and their nuclei and the gaseous components of the stratosphere, modification of the equation for vapor pressure of the H_2SO_4–H_2O system in the gaseous state, and development of a scheme of homogeneous heteromolecular and ionic nucleation for an understanding of the role of similar nucleation in unusual conditions in the stratosphere.

In view of the above, carrying out more comprehensive observations than those undertaken so far is necessary. Primarily, the measurement of total ratio of the mixture of aerosol particles is necessary in order to reveal the mechanism of particle formation. Aerosol masses may prove to be of special interest in the study of homogeneous and ionic nucleation. For testing the conclusion in regard to evaporation of sulfuric acid drops at altitudes above 25 km, measurements of the mass ratio of the mixture and of coarse particles are of great importance. The measurements (possibly from satellites) of concentration of coarse particles above cold winter and warm summer poles may also serve the same purpose.

The investigations of vertical profiles of concentration of such minor components as ammonia, silicon, iron, sodium, carbon, chlorine and water vapor are important for an understanding of the sources of aerosol nuclei together with their formation from the gaseous phase and the formation of sulfuric acid. Collection of data on latitudinal variations of the aerosol microstructure is also of great interest. Special measurements of the microstructure of finely dispersed aerosols are required for testing the concept of homogeneous nucleation. From the standpoint of investigations of homogeneous nucleation and growth of nuclei, the determination of the fraction of droplet volume, occupied by the nucleus, as a function of the size of the droplet is also of considerable interest. Measurements of the size and composition of condensation nuclei near the tropopause are required. After volcanic eruptions it should have been possible to observe variations in the ratio between sulfuric acid droplets and particles of ash from the viewpoint of their total concentration and concentration of coarse particles as well as concentration of particles without nuclei. Observations of releases of gases of anthropogenic origin (high altitude aviation and tropospheric sources of sulfur dioxide), anthropogenic nuclei of condensation and their possible effect on stratospheric aerosol are of extreme importance.

We shall now discuss some results of theoretical investigations show-ing the effect of stratospheric aerosol on the transfer of shortwave and long-wave radiation. Luther [66–70] carried out a series of exhaustive investigations on this problem. The author proposed a model of transfer of shortwave radiation in a cloudless plane-parallel atmosphere on the basis of application of the method for calculating shortwave radiation fluxes developed by Braslau and Dave [26, 27]. The model takes into account molecular and aerosol scattering (in the stratosphere), as well as molecular (water vapor, carbon dioxide, ozone, oxygen) and aerosol ab-sorption.

2.1. Albedo of the system. In connection with the assessment of ac-curacy of approximate methods of calculation of the albedo of the earth's surface-atmosphere system, Luther [70] compared the values of total al-bedo obtained: 1) by integrating the calculated spectral albedo for the 0.285–2.5 μm band, 2) from the data on albedo on the 0.55 μm wave-length. In both cases, the method of computation of albedo was the same, while calculations have been carried out for zenith angles of 30, 60 and 80° of the sun, and an albedo of the underlying surface of 0.0, 0.1, 0.15, 0.25, 0.50 and 0.80 for the model of a tropical atmosphere, with and with-out aerosol. The calculations showed that the 'single-wave' method should not be considered reliable enough in the case under consideration, be-cause discrepancies may emerge even in the sign of variability of the albedo obtained by accurate calculations. Possibly, a better agreement may be obtained by changing over to another wavelength.

In [67], the results of computation of albedo for the surface-atmo-sphere system in the 0.285–2.5 μm wave band have been discussed. The wave band for three zenith angles of the sun (30°, 60° and 80°) and five values of underlying surface albedo (0.0, 0.15, 0.25, 0.5 and 0.8) has been divided into 83 spectral intervals. The purpose of analysis was to assess the impact of variation of stratospheric aerosol content in the 18–22 km layer and that of ozone ($\Delta O_3/O_3$) in the 20–30 km layer on the global albedo in a cloudless sky for average surface conditions (altitude of the sun—60°). The values under consideration have been obtained as a mean for the albedo of the underlying surface equal to 0.1 to 0.5.

In the first column of Table III.7, the 'reference' absolute values of radiation fluxes, albedo corresponding to the variations of radiation fluxes (W/m²) and albedo presented in subsequent columns are indicated. In the case under consideration, both a decrease in ozone content and a growth of optical thickness of aerosol result in an increase of albedo for the earth's surface-atmosphere system (cooling of climate). However, the redistribution of energy between the stratosphere and the troposphere takes place differently if depletion of ozone causes attenuation in absorption by the stratosphere, and absorption by the troposphere increases. In the case of an increase in dust content of the stratosphere, there exists a reverse situation (effect of warming of the stratosphere and cooling of the troposphere, whereas, for the earth's surface-atmosphere system, the second effect is roughly twofold). It is natural that in actual conditions the proportion of the effects discussed may greatly vary depending on the properties (and the quantity) of aerosol, the quantity of ozone, the zenith distance of the sun and other factors.

Table III.7. Effect of decrease of ozone content ($\Delta O_3/O_3$) and increase of optical thickness of aerosol ($\Delta\tau$) in the stratosphere on radiation fluxes and albedo of system

Radiational characteristics	Radiation fluxes	$\Delta O_3/O_3$		$\Delta\tau$	
		−0.052	−0.13	0.005	0.02
Variation of outgoing shortwave radiation, W/m^2	481.9	−0.4	−1.1	−0.3	−1.3
Variation of albedo of system	0.2819	0.0006	0.0016	0.0005	0.0019
Variation of radiation absorbed by stratosphere, W/m^2	21.4	−0.9	−2.3	−0.3	1.1
Variation of downward fluxes of shortwave radiation at the level of tropopause, W/m^2	460.5	0.5	1.2	−0.6	−2.4
Variation of radiation absorbed by troposphere, W/m^2	92.1	0.1	0.2	−0.1	−0.3
Variation of radiation absorbed by earth's surface, W/m^2	368.4	0.4	1.0	−0.5	−2.1

The data of Table III.7 indicate only the estimates of effects related with the transfer of shortwave radiation. Calculations show that the contribution of long-wave radiation is also significant. Thus, for example, a decrease in ozone content by 50% leads to a rise in the earth's surface temperature by 1.5 °C due to an increase in the absorption of solar radiation, but to a decrease of 1.0 °C due to radiant heat exchange. Further on a comparative assessment of the radiational effects of aerosol in the shortwave and long-wave bands will be presented.

In [24], calculations of the effect of sulfate-bearing stratospheric aero-
232 sol on the albedo of the earth's surface-atmosphere system based on the application of a two-stream approximation have been carried out. Computations were undertaken for various values of zenith distance of the sun, the albedo of single scattering with values of 1.0, 0.9, 0.8, 0.7, 0.6 and concentrations of stratospheric aerosol ρ equal to 0.0, 0.2, 0.3, 0.4, 0.8, 1.6 and 3.2 μg/m^3. It was assumed that the variation in the optical

layer of stratospheric aerosol $\Delta\tau=0.038\ \rho$. The albedo of the underlying surface constitutes 0.1, 0.3, 0.9. If we assume that the concentration of stratospheric aerosol of natural origin is 0.6 $\mu g/m^3$, scattering from aerosols alone causes a drop in temperature by 0.7 °C. The effect of aerosol pollution created by 2000 supersonic *Concorde* aircraft on temperature would not exceed 0.13 °C and the effect from 300 aircraft would be negligibly small.

Herman et al. [52] carried out calculations to assess the effect of stratospheric aerosol with different optical properties on the albedo of the earth-atmosphere system and distribution of temperature in the conditions of radiative equilibrium. The microstructure of stratospheric aerosol was described by a logarithmic normal distribution and that of tropospheric aerosol by Young's formula. It was assumed that aerosol particles with a density of 2.0 g/cm^3, are uniformly distributed in the layer of 10 km thickness (15–25 km).

Calculations of the optical thickness of the aerosol layer led to the conclusion that this closely depends on the parameters of the microstructure. In calculations relating to a characteristic wavelength of 500 nm, the Rayleigh ratio and background aerosol optical thicknesses were taken as 0.145 and 0.100, respectively, and the index of refraction of particles as 1.54. It was assumed that tropospheric aerosol is purely scattering in nature, and in the case of stratospheric aerosol various values of the albedo of single scattering ω_0 in the 1.0–0.6 μm range were assumed. Further, stratospheric aerosol consisting of droplets of a 75% water solution of sulfuric acid was considered. The albedo of the underlying surface A_s accepts values of 0.05, 0.1, 0.3, 0.5, 0.7 and 0.9, the zenith distance of the sun ϑ_0 varies in the limits 5–85° (in steps of 10°), and the concentration of contaminating aerosol in the 15–25 km layer is equal to 0.8, 1.6, 2.4 and 3.2 $\mu g/m^3$.

The authors of [52] analyzed the relationship between the flux of reflected (outgoing shortwave) radiation and various parameters and conditions of decrease or increase of reflected radiation in the case of an increase in aerosol concentration. Thus, for example, for $A_s=0.1$ and $\vartheta_0=15°$, an increase in aerosol concentration causes an increase in reflected radiation (cooling effect), when $\omega_0>0.88$. On the contrary, for $\omega_0<0.88$, the situation is exactly the opposite. Since the values $\omega_0\approx0.8$–0.9, correspond to frequently occurring values of the imaginary part of the complex index of refraction (absorption index), equal to 0.01, hence for $A_s=0.1$, the effect of cooling or warming is highly dependent on the absorption index and the zenith angle of the sun. For a higher albedo of the surface ($A_s=0.9$), a warming effect occurs even in the case of non-absorbent particles.

Calculations of the annual trend of reflected radiation in nine latitudinal steps of 10° each in the northern hemisphere for ω_0 equal to 1.0, 0.9 and 0.7 enabled us to analyze the effects of warming or cooling under actual conditions. For $\omega_0 = 1.0$, south of 40°N, an increase in reflected radiation for an aerosol concentration of 0.8 $\mu g/m^3$ indicates a poor dependence on latitude and time of the year, and constitutes about 0.4%. The variability of reflected radiation at higher latitudes is more distinct, attaining 0.7% in January and February. A similar situation exists for the albedo of single scattering equal to 0.9 and 0.7, but the sign of effect changes (warming) with a maximum decrease of reflected radiation at higher latitudes of up to 1.8% for $\omega_0 = 0.9$ and 6% for $\omega_0 = 0.7$.

234   Estimates of changes of mean temperature in the entire thickness of the atmosphere under the assumption of radiative equilibrium are presented in [52]. If $\omega_0 = 0.7$, the maximum warming observed in May and June can reach up to 3.5 K/day. The calculations pertaining to the case of sulfur-bearing aerosol and computations carried out for the complete spectrum of solar radiation (0.4–2.5 μm) also revealed that the most significant changes of reflected radiation take place at higher latitudes during early autumn and late winter and can be up to 1.4%. At lower latitudes these variations as a rule do not exceed 0.3% in the course of a year. The calculations of temperature variations for conditions of radiative equilibrium have revealed a cooling of up to 0.5 K/day in polar regions.

Hummel [53] carried out approximate calculations of the effect of the mesospheric aerosol layer consisting of spherical ice particles, on the albedo of the earth's surface-atmosphere system and manifest in view of this climatic effect a warming (or cooling) for decrease (or increase) of the albedo of the system. It was assumed that the aerosol layer of 3 km thickness is situated at an altitude of 85 km and covers the hemisphere experiencing summer in the latitudinal zone above 75°. The computations were carried out with the consideration of growth and subsequent dissipation of the layer. Four models of aerosol microstructure encompassing the 0.01–0.17 μm range of radii have been considered. The albedo of the underlying surface was taken as 0.7.

The calculations revealed the decisive effect of particles with a radius smaller than 0.09 μm on the albedo in creating a warming effect which, in turn, compensates for the effect of cooling as a result of the influence of coarser particles. Precisely because of this, the earlier calculations carried out for single-scattered aerosol with a radius of particles of 0.17 μm, led to an erroneous conclusion on the cooling effect caused by mesospheric aerosol. At the same time, the diffusion property of reflected radiation is of considerable importance. The warming effect occurring just after the formation of the aerosol layer intensifies with the growth of the layer and attains its maximum value on the 119th day, after which the value starts declining. Although the absolute value of the effect is not large (maximum attenuation of radiation by the aerosol layer is about 1%), it may all the same, influence the climate.

2.2. Absorbed radiation. Let us now discuss some results obtained by Luther [66] from calculations of the variability of radiant heat influx into the stratosphere because of absorption of shortwave radiation by aerosol under conditions when the composition, microstructure and optical properties of stratospheric aerosol remain constant, but the zenith angle of the sun and albedo of the underlying surface may vary. The
235 basis of these calculations is the 160-level model of the atmosphere with a resolution along the vertical axis equal to 1 km in the 0 to 25 km layer and 5 km in the 25–50 km layer. The vertical profiles of pressure, temperature, water vapor and ozone correspond to the standard atmosphere of the tropics. For the microstructure of aerosol the haze model H, proposed by Deirmendjian was considered. The particle concentration in the 18–22 km layer is 100 cm^{-3} and linearly decreases to zero at the levels of 17 and 23 km. In such a case the total aerosol content is 5×10^7 particles/cm^2, equivalent to 2.36 μg/cm^2 or 12×10^6 t for the global atmosphere. The complex index of refraction was taken as 1.45–0.005i. The optical thickness of the aerosol layer at the 0.55 μm wavelength is 0.109, which exceeds almost by five times the normal optical thickness of the stratosphere according to Alterman and is half the increase in optical thickness of the stratosphere after the eruption of Mount Agung.

Calculations have been carried out without consideration of the effect of cloudiness and tropospheric aerosol. The value of the radiant heat influx on account of absorption of shortwave radiation (radiative fluctuation of temperature) has been calculated for zenith angles of the sun of 30°, 60° and 80°, and an albedo of the underlying surface equal to 0.0, 0.25, 0.5 and 0.8. An analysis of the results of computations shows that the maximum variations of absorbed solar radiation take place near the upper boundary of the aerosol layer. Thus, for example, for a zenith angle of 30°, the increase in fluctuations of temperature due to radiation at the level of 21.5 km is 1.0 °C/day when the surface albedo equals 0.0, and 1.8 °C/day in case of an albedo value of 0.8. For a zenith angle of 80° the corresponding values are 1.0 and 1.1 °C/day. The appearance of an aerosol layer would increase the shortwave radiation absorbed in the 18–22 km layer by more than two times. For small optical thicknesses (less than 0.1) the variation of radiant heat influx, in the first approximation, is related linearly to the aerosol content.

2.3. Effect of aerosol on long-wave radiation. The assessment of the effect of stratospheric aerosol on long-wave radiation fluxes and radiant heat exchange is of considerable interest. Luther [68] obtained estimates of the aerosol impact on the long-wave radiation fluxes at the upper and lower boundaries of the aerosol layer in summer and winter at various

224

latitudes on the assumption that stratospheric aerosol consists of droplets of a 75% sulfuric acid solution. The microstructure of aerosol has been assumed by him from the Deirmendjian has model H and the content of stratospheric aerosol in the 18–22 km layer was assumed to have increased by 0.1 $\mu g/cm^2$, which corresponds to an average increase of optical thickness of the layer by 0.005 for the 0.55 μm wavelength and in the 236 infrared region of the spectrum (6–50 μm) by 0.00025. It is accepted that in the infrared region of the spectrum, the particles appear to have a purely absorbent nature.

Calculations on the basis of data on stratification of the atmosphere in the tropical, temperate and subtropical latitudes during winter and summer enabled us to obtain the value indicated in Table III.8 (where T is the temperature at a latitude of 20 km and T_e the effective temperature assumed equal to the air temperature at an altitude of 6 km or about 500 mb). Apparently, an increase in the downward radiation flux in the troposphere (at the lower boundary of the aerosol layer) is observed, being most significantly expressed during summer and at higher latitudes, i.e. thermal radiation of stratospheric aerosol promotes a heating of the troposphere. On the other hand, a decrease in the downward radiation flux over the aerosol layer is poorly expressed at higher latitudes and demonstrates the decrease of outgoing radiation under the influence of aerosol. This decrease of outgoing radiation to a certain degree compensates for the decrease in absorbed shortwave radiation caused by stratospheric aerosol (Table III.8). The aerosol layer is warmed due to a radiant heat exchange at lower latitudes and cools down at higher latitudes. Though all variations of radiation fluxes are small (global mean value of outgoing radiation is 233 W/cm^2), they can be significant; however, for clarification of this issue their comparison with corresponding variations of fluxes and shortwave radiation influxes is essential.

On the basis of the application of a simple two-layer (troposphere +stratosphere) model of the radiation budget of the earth's surface-

Table III.8 Variation of long-wave radiation fluxes above ($\Delta F\uparrow$) and below ($\Delta F\downarrow$) aerosol layer and radiant heat influx Δq in it

Quantity	Tropics	Temperate latitudes		Subtropical latitudes	
		summer	winter	summer	winter
$\Delta F\uparrow$ W/m²	0.43	0.53	0.51	0.50	0.49
$\Delta F\downarrow$ W/m²	−0.71	−0.56	−0.32	−0.36	−0.21
Δq W/m²	0.28	−0.03	−0.19	−0.24	−0.28
T K	207.0	218.0	215.2	225.0	214.1
T_e K	264.0	261.0	243.7	253.0	234.1

237 atmosphere system, while assuming the presence of radiative equilibrium for the system as a whole and the stratosphere, Coakley and Grams [38] undertook calculations of the relative variation of outgoing thermal radiation $\Delta F_\infty/F_\infty$ at the level of the upper boundary of the stratospheric aerosol layer for an increased content of aerosol in the stratosphere. If we introduce characteristics of molecular absorption averaged over the spectrum then $\Delta F_\infty/F_\infty$ will depend only on the optical properties of aerosol particles. The use of the equation of thermal budget enables us to express $\Delta F_\infty/F_\infty$ in terms of variations in the mean global temperature of the earth's surface.

The heat-budget models of climate reveal that a decrease in outgoing radiation by 1%, is accompanied by a lowering of the surface temperature by approximately 1 K. On the basis of this result the effect of fluctuations of the ratio $\Delta F_\infty/F_\infty$ caused by stratospheric aerosol on the mean global temperature of the underlying surface can be assessed, assuming that variations in the latter do not affect the characteristics which determine the optical properties of the troposphere (cloud cover and surface albedo).

Calculations of variations in $\Delta F_\infty/F_\infty$ have been undertaken for spherical particles of aerosol. The particles acquire different values of the complex index of refraction in the visible ($1.5–0.0i$ and $1.5–0.01i$) and infrared ($1.5–0.1i$ and $1.5–0.5i$) regions of the spectrum. The results of calculations have been presented in the form of a ratio $\Delta F_\infty/F_\infty$, with the radius 'r' of the particles for a given mass concentration of stratospheric aerosol equal to $1\mu g/m^3$ (for a density of particles of 2 g/cm^3).

It has been observed that very fine ($r \gtrsim 0.05\mu m$) and coarse ($r \lesssim 1.0\mu m$) particles affect the long-wave radiative transfer more significantly than the shortwave radiation, a fact that determines the heating effect caused by such particles on the earth's surface. On the other hand, a situation opposite to this accompanied with cooling prevails in the case of particles of intermediate radii ($0.1–1.0$ μm).

In [38] it has been shown that the variation in surface temperature is maximum in the case of single-scattered aerosol, consisting of particles with a radius of about 0.2 μm. The actual fluctuation of temperature should be less than that observed because: 1) stratospheric aerosol is not single-scattered and 2) a higher time constant (thermal inertia) of the earth's surface-atmosphere system, apparently, modulates the temperature variations of the surface to a considerable extent. If these factors are not taken into account, then the dust content of the stratosphere, equivalent to the level of consequences of eruption of Mount Agung, should have caused a lowering of the mean global temperature of the earth's surface by 0.8 K.

238 If the obtained dependence of $\Delta F_\infty/F_\infty$ on the radius of aerosol

particles is used, then considering that a major fraction of current stratospheric aerosol consists of particles with a radius of between 0.1–1.0 μm, it should be noted that such aerosol causes a cooling of the earth's surface. Here it is important to mention the inaccuracy of the often made assumption about the equivalence of the effect of the aerosol layer and variation of solar constant on climate (especially in the case of particles with a radius above the 0.1–1.0 μm range). Such an equivalence occurs only when aerosol particles do not affect the long-wave radiative transfer and possess a nonabsorbent property in the shortwave spectral band.

The conclusions obtained in [38] are independent of the details of the method of computation. The proportionality between $\Delta F_\infty/F_\infty$ and the surface temperature alone are very important in a case where the proportionality is positive. Such proportionality is confirmed by satellite measurements of the radiation budget of the earth, at least as applied to average values of hemispheric temperature of the surface for time scales of the order of seasonal variations. It should be viewed as a confirmation of this conclusion that an increase in stratospheric aerosol content would entail a cooling near the earth's surface.

2.4. Radiation budget. We shall carry out a detailed comparison of results of computations, undertaken by Luther [69], of long-wave and shortwave radiation fluxes with the aim of assessing the effect of tropospheric aerosol in the 18–22 km layer on the radiant heat influx and radiation budget of the earth's surface-atmosphere system in the conditions of a tropical atmosphere in the absence of clouds. The values of long-wave radiation fluxes have been obtained from the computational data of Ellingson carried out with the consideration of water vapor, carbon dioxide, ozone, methane (7.66 μm wavelength) and nitrogen oxide (7.78 μm). These calculations were undertaken on a spectral basis (100 intervals in the range of 3.55 μm to ∞) with the consideration of the effect on the transfer of radiation due to multiple scattering by aerosol.

Calculations of shortwave radiation fluxes are based on the above method. Detailed computations have been carried out for 83 spectral intervals in the 0.285–2.5 μm band and the 500-level model of the atmosphere. It has been assumed that stratospheric aerosol consists of droplets of a 75% sulfuric acid solution (complex index of refraction $m = 1.45–0.005i$) and its microstructure corresponds to the haze model H or model M of Deirmendjian (the first of these models is considered to be a realistic one). In both cases, the total aerosol content in the stratosphere is the same, at a value of 2.36 μg/cm^2. This corresponds to 5×10^7 particles/cm^2 in the case of the haze model H and 3.175×10^6 particles/cm^2 in the case of haze model M for which the predominance of coarser particles is characteristic. The effect of tropospheric aerosol is not taken into consideration. In case of a small optical thickness (less than 0.1), variations in shortwave and long-wave fluxes linearly depend on the total ozone content, so that extrapolation of the obtained values provides reliable results.

Results of computation of long-wave radiation fluxes are presented in Table III.9, and these demonstrate that the effect of aerosol micro-structure on the transfer of long-wave radiation is practically negligible. The presence of stratospheric aerosol in the case considered promotes a warming of the troposphere and the stratosphere (increase in the contents of stratospheric aerosol should cause an increase in warming).

Table III. 9. Total upward and downward long-wave radiation fluxes at the lower (17 km) and upper (23 km) boundaries of aerosol layer (W/m²)

Models of aerosol	Upward flux	Downward flux	Long-wave radiation budget	Change of radiation budget
Altitude 23 km				
Without aerosol	288.9	9.1	—279.8	
Haze H	287.4	9.1	—278.3	+1.5
Haze M	287.4	9.1	—278.3	+1.5
Altitude 17 km				
Without aerosol	290.9	10.8	—280.1	
Haze H	290.9	11.6	—279.3	+0.8
Haze M	290.9	11.6	—279.3	+0.8

Calculations of shortwave radiation fluxes showed that mean diurnal variations of the shortwave budget at the boundaries of the aerosol layer closely depend on the albedo of the underlying surface and the micro-structure of aerosol. At a sufficiently high albedo of the surface the presence of stratospheric aerosol promotes a warming of the earth's surface-atmosphere system below the upper boundary of the aerosol layer. For a low albedo the effect of cooling caused by an increase in the albedo of the system is evident. In the case of the haze model H, the transition from total cooling (with the consideration of shortwave and long-wave radiation) to warming of the 0–23 km layer takes place for a surface albedo of more than 0.35. If the average value of the global albedo is considered to be about 0.3, then it may be concluded that the effect of stratospheric aerosol on the radiation budget of the system due to shortwave and long-wave radiation is approximately equal in value, but opposite in sign. The calculations relating to the lower boundary of the aerosol layer (17 km) show that for any value of albedo the strato-spheric aerosol causes an attenuation of tropospheric insolation much more precisely expressed in the case of the haze model H. Hence, cooling of the troposphere exceeds the warming caused by long-wave radiation by 2–5 times. Therefore, the resultant effect of stratospheric aerosol is nothing but a cooling of the troposphere. The absorption of shortwave radiation by the layer of aerosol itself causes this warming, exceeding in value by 2–4 times the warming due to long-wave radiation.

Insofar as a variation in the shortwave budget caused by stratospheric aerosol is more strongly manifest at the level of the lower boundary of the aerosol layer, this means that even in a case where stratospheric aerosol does not affect the global albedo, its presence can significantly influence the radiation budget of the underlying surface-troposphere system and consequently the surface temperature. Naturally the continuation of such calculations is necessary to encompass much more varied conditions (latitude, time of year, etc.). However, it can be safely assumed that stratospheric aerosol promotes a cooling during summer and warming during winter at higher latitudes [69].

Pollack et al. [92] carried out computations of variations in the thermal budget of the earth caused by stratospheric aerosol with the aim of estimating variations in the mean global temperature ΔT of the earth's surface caused by these variations (for a five-point cloudiness and an albedo of the earth's surface at 10%). Since calculations do not take into account the number of inverse relations (for example, the inverse relation caused by variations of snow and ice cover), the consideration of which enabled us to establish that they lead to an increase of ΔT by approximately 4 times, the corresponding correction has been introduced in the calculations for time scales of more than 10^3 years. For shorter time scales the value of the correction F_1 has been reduced. Keeping in mind that all calculations have been undertaken for a steady state model, the authors of [92] have taken into consideration the effect of the nonsteady nature by introducing a correction factor $F_2 = P/2t$, where P is the period of existence of a stable perturbation of climate and t the characteristic reaction time for a continental atmosphere-ocean system.

This method has been used for calculation of ΔT, caused by volcanic eruptions of the 'explosive' type observed in the past. Table III.10 contains a summary of the results obtained. In the first two columns is given the data on the effects of single eruptions (the first column signifies the effect of warming of the lower stratosphere primarily caused by absorption of upward long-wave radiation by stratospheric aerosol), and the remaining results characterize the cumulative effect of multiple eruptions and are found to lie (from top to bottom) in the order of increasing duration of climatic variations ($\Delta\tau$ is the variation in optical thickness of the stratosphere as a result of eruption).

A comparison between computed and observed values can serve as a

Table III.10. Comparison of observed and computed variations of mean global temperature of earth's surface caused by impact of aerosol of volcanic origin

Variation of temperature, K		Hemisphere	Duration	Type of aerosol	Parameters		
observed	computed				$\Delta\tau$	F_1	F_2
0.1	0.1	Southern	First month after eruption of Mount Agung, 1963	Silicate dust	0.1	—	—
−0.3	−0.2	Northern	Mean for two-year period after eruption	Sulfuric acid	0.1–0.0	0.25	1
0.6	0.5	—do—	1935–1945 as compared with 1880–1890	—do—	0.0–0.07	1	1
−0.3	−0.15	—do—	1965–1970 as compared with 1935–1945	—do—	0.02–0.0	1	1
0.5	0.3–0.6	—do—	1935–1945 as compared with 1850–1915	—do—	0–0.05	1	1–2
0.4	0.7–7	Global	Contemporary period as compared with 12000–25000 years ago	—do—	0–0.2	1	4

criterion of reliability of the calculations. According to the authors of [92], a decrease in the intensity of volcanic activity from the end of the nineteenth century till the beginning of 1940 should be considered as the main reason for a global warming observed during this period. The role of an increasing carbon-dioxide content at that time was less pronounced, but during several subsequent decades, the situation could have been reversed, if only the effect of anthropogenic sources making the atmosphere dust laden had not existed. High volcanic activity should have emerged as an important factor of a 'minor ice age' (1450–1715). The same can also be assumed regarding the Burms period of freezing (12000–25000 years ago). It should be emphasized, however, that volcanic eruptions were an important though not the only factor responsible for climatic variations in the past.

The application of a similar technique for forecasting ΔT, evolved due to the effect of supersonic aviation and space transport vehicles in the course of the forthcoming few decades shows that an increase of stratospheric aerosol content caused by these factors will be less than the increase occurring due to a single weak volcanic eruption. In this case all calculated values of ΔT are less than 0.1 K, i.e. the assumed effect of supersonic aviation and space transport vehicles related to variations in the stratospheric aerosol content should not be considered significant. A drop in temperature by 0.1 K is possible only in a case where 70 space transport vehicles (toward the year 2000 not even one launch a day is planned), and 5000 supersonic aircraft (250–1500 are anticipated) are launched daily. The effect of anthropogenic aerosol on stratospheric ozone would also be negligibly small.

Mugnai et al. [73] carried out calculations of the thermal budget and temperature of aerosol particles in the stratosphere, as well as the impact of aerosol on the thermal regime of the stratosphere. The study of the thermal budget included consideration of absorbed solar radiation and heat exchanges with the surrounding air due to radiational conduction. The main aim of the calculations continuing the investigations earlier undertaken is to simulate conditions in the stratosphere, observed after the eruption of Mount Agung in 1963. The comparison of observed temperature fields during January, 1973, and January, 1974, at an altitude of 19 km enabled us to observe a considerable rise in temperature in the 0–20°N belt with a maximum rise (up to 6–8 K) above Africa, Guinea and South America. These maximums should be attributed to an increased albedo of the underlying surface, which causes an increase in solar radiation absorbed by aerosol. Such a conclusion is confirmed by a quantitative similarity of fields of isolines characterizing the increase in temperature and albedo of the earth's surface-atmosphere system.

In [73] the calculations of global distribution (in the region from 40°S and 40°N) of the mean daily rates of radiant heating caused by strato-spheric aerosol at an altitude of 19 km, have been carried out. The said aerosol consists of particles of volcanic ash and droplets of a contami-nated 75% aqueous solution of sulfuric acid. Calculations were under-taken with consideration of the data on the spatial distribution of aerosol concentration on the 300th day after eruption of Mount Agung (mid-January, 1964). The microstructure of aerosol with a radius in the range of 0.03–3 μm has been accepted in accordance with assumptions made by Young. The results of computations reveal a distinct correlation between the global distribution of radiant heating attaining 0.4 K/day in the maximal zones and the aforementioned temperature rise at the level of 19 km. Comparison of the maximum rates of radiant heating (0.4 K/day) and temperature rise (8 K/day) lead to the conclusion that the time of radiative relaxation at an altitude of 19 km is about 20 days.

The results of computations of the effect of ice particles (silver clouds) on the thermal regime of the mesosphere by calculating the equilibrium temperature of ice particles in the upper atmosphere at an altitude of 68–95 km, where silver iodide clouds can be formed, showed that de-pending on the altitude above sea level and size of the particles, their temperature may be higher than the temperature of the surrounding air by approximately 50 K.

An analysis of the results obtained enables us to conclude that, during summer, the range of altitudes at which the formation of ice clouds is possible significantly decreases at lower latitudes. During winter, the conditions for the formation of silver iodide clouds at all latitudes are less favorable than during summer. The most interesting conclusion related to the fact that the formation of ice particles in the mesosphere in January at 60°N is impossible. The conclusions obtained agree with the results of observations of the annual trend of silver iodide clouds at higher latitudes.

Conclusion

244   The leading role of radiation factors of climate generates considerable interest toward the investigations of the radiational component of the energetics of the atmosphere and the underlying surface. If the problems related to the determination of radiant heat influx caused by molecular absorption and radiant heat exchange (in the conditions of an aerosol-free atmosphere) can be considered more or less solved, then the problem concerning the impact of aerosol on radiative transfer and the corresponding climatic effects remains, in many aspects, unclear. This concerns radiative transfer both in the presence of water aerosol (clouds) as well as dry aerosol (haze).

Apart from the problem of cloudiness (here the main problem consists in the search for a statistical approach to the solution of problems relating to radiative transfer in the presence of multilevel partial cloudiness), we shall emphasize that in the investigations of the climatic effect of aerosol the main barrier is the absence of adequate data on global aerosol. In this connection the carrying out of an extensive program of investigations covering the spatial (three-dimensional) distribution and temporal variability of aerosol concentration, its chemical composition, microstructure, form and complex index of refraction of particles is highly important. For this purpose both direct and indirect methods of measurements (laser soundings, spatial remote sensing) should be used.

The important problem in the analysis of data from such measurements is to determine the ratio between aerosol of natural and industrial origin.

The development of a theory explaining the formation of global tropospheric and stratospheric aerosol no doubt, has exceptional importance. Any theory, based on observational data, could have become the 'aerosol block' in the theory of climate. Only such an approach would open prospects of an adequate theoretical assessment of the climate-form-
245 ing role of atmospheric aerosol. In this connection, the development of a theory covering the global cycles of carbon, sulfur, nitrogen and a few other components serving as sources of aerosol becomes important.

The effect of atmospheric aerosol can manifest itself in different ways. These are the impact of aerosol on cloudiness, which in its turn causes variations of the radiation regime of the atmosphere, and the direct impact of aerosol on shortwave and long-wave radiative transfer. The first of the aforementioned aspects appears to be the most significant, since clouds are instrumental in controlling the radiation regime of the atmosphere and the underlying surface. Even small (of the order of a few percent) variations in the amount of clouds or their optical properties can affect the climate considerably more than large variations in aerosol as a result of dust content.

Till now, only hypothetical conjectures about the possible effect of aerosol on clouds (effect of increased concentration of condensational nuclei on the conditions of formation of clouds, variation of optical properties as a result of dissolution or accretion of aerosol substance by water droplets, etc.) have been presented. Therefore, investigation of the role of aerosol in the formation and evolution of a cloud cover is of primary importance. Our information on optical characteristics of clouds is far from complete (optical thickness, albedo, scattering index). Meanwhile, it is evident that only precision measurements of the optical and microphysical characteristics as well as chemical composition of droplets of clouds (along with the data on dry aerosol) would enable us to solve the problem of interaction between aerosol and cloudiness. Experiments in this connection, not only under natural, but in laboratory conditions, are essential.

As far as the direct effect of aerosol on the transfer of radiation is concerned, the chief ways of solving this problem are connected with further development of the radiative transfer theory and results of complicated experiments enabling us to obtain simultaneously data on aerosol (concentration, microstructure, optical properties) and also important radiational characteristics (albedo of the earth's surface-atmosphere system, in shortwave radiation absorbed in various layers of the atmosphere, radiant heat exchange, etc.). It is important that such complex 'aerosol-radiation' experiments be undertaken in different zones of the globe (tropical oceans, deserts, plains, ice shelves of the Arctic and the Antarctic, forests of tropical and temperate latitudes, etc.).

246 At the same time stratospheric aerosol merits special attention both from the viewpoint of investigation of the physicochemical processes causing its formation as also in respect to its influence on the radiation regime of the stratosphere, the troposphere and the underlying surface.

The problems mentioned above, and many other issues related with 'aerosol and climate', determine the immense significance of the Global Atmospheric Research Progamme (GARP), so essential for the solution of an entire set of problems that need to be resolved.

Bibliography

CHAPTER I

1. Barashkova, E.P. and I.P. Vinogradova. 1972. Geographicheskoe raspredelnie voskhodyashchego dlinnovolnovogo izlucheniya pri bezo-blachnoi atmosfere (Geographical distribution of upward long-wave radiation in cloudless atmosphere). *Trudy GGO*, vol. 279, p. 11–23.
2. Bogorodskii, V.V., A.I. Kozlov and L.T. Tuchkov. 1977. Radioteplovoe izluchenie zemnykh pokrovov (Radiothermal radiation of earth's covers). Gidrometeoizdat, Leningrad. 224 p.
3. Bogorodskii, V.V., E.A. Martynova and V.A. Spitsyn. 1977. Issledo-vanie formirovaniya sobstvennogo teplovogo izlucheniya snezhno-ledy-anogo pokrova arkticheskikh morei primenitel'no k zadacham IK radiometrii (Investigation of formation of inherent thermal radiation of snow and ice cover of Arctic Sea as applicable to the problems of infrared radiometry). *Trudy GGO*, vol. 399, p. 87–114.
4. Bogorodskii, V.V. and A.I. Paramonov. 1977. IK diagnostika vozra-stnykh gradatsii dreifuyushchikh l'dov i issledovanie dinamiki pover-khnostnykh temperatur vod v vostochnoi chasti Arktiki (Infrared diagnostics for age gradation of floating ice and investigation of surface temperature dynamics of waters in the northern Arctic). *Trudy GGO*, vol. 399, p. 115–127.
5. Borisenkov, E.P. and V.N. Priemov. 1976. Energeticheskaya otsenka klimaticheskikh trendov poslednego stoletiya (Assessment of climatic trends of the last century). *Pis'ma v astronom. zhurnal*, vol. 2, No. 1, p. 44–49.
6. Budyko, M.I. and K.Ya. Vinnikov. 1976. Global'noe poteplenie (Global heating). *Meteorologiya i Gidrologiya*, No. 7, p. 16–26.
7. Buchinskii, I.E. 1977. O chem govoryat mnogoletnie nablyudeniya v Voroshilovgrade (What the observations of several years at Voroshi-lovgrad reveal). In *Chelovek i stikhiya*. Gidrometeoizdat, Leningrad. p. 51–52.
8. Bushuev, A.V. and N.A. Volkov. 1969. Meteorologicheskie iskysst-

vennye sputniki Zemli kak sredstvo nablyudeniya za l'dami (The meteorological earth satellite as an observer of ice). *Problemy Arktiki i Antarktiki*, vol. 33, p. 5–12.

9. Volkov, N.A. and V.F. Zakharov. 1977. E'volyutsiya ledyanogo pokrova v Arktike v svyazi s izmeneniyami klimata (Evolution of ice cover in the Arctic due to changes in climate). *Meteorologiya i Gidrologiya*, p. 47–55.

10. Kudritski, D.M. (Ed.). 1977. Voprosy metodiki distantsionnogo izucheniya vodnykh resursov (Problems of methods of remote study of water resources). *Trudy GGI*, vol. 232, p. 1–169.

11. Gedeonov, A.D. 1973. Izmeneniya temperatury vozduxa na severnom polusharii za 90 let (Temperature changes in the northern hemisphere during the last 90 years). Gidrometeoizdat, Leningrad. 146 p.

12. Grigor'eva, A.S. and L.A. Strokina. Kolebaniya v khode temperatyr vysokikh shirot severnogo polushariya (Temperature fluctuations at higher latitudes of the northern hemisphere). *Trudy GGI*, vol. 247, p. 114–118.

13. Gruza, G.V., L.K. Kleshchenko and E'.Ya. Ran'kova. 1977. Ob izmeneniyakh temperatury vozdukha i osadkov na territorii SSSR za period instrumental'nykh nablyudenii (Temperature variation of air and rains in the USSR during a period of instrumental observations). *Meteorologiya i Gidrologiya*, No. 1, p. 3–25.

14. Bushueb, A.V. and N.A. Volkov (Ed.). 1977. Distantsionnye izmereniya parametrov ledyanogo pokrova (Remote measurements of ice-cover parameters). *Trudy AANII*, vol. 343, p. 1–154.

15. Borisenkov, E.P. et al. 1977. Dolgoperiodnye kolebaniya pogody u klimata i ikh prognozirovanie (Long term variations of weather and climate and their forecasting). In *Sovrem. fundament i prikladn. issled.* Gl. geophiz. Obs. Gidrometeoizdat, Leningrad. p. 40–52.

16. Drozdov, O.A. 1976. Struktura kolebanii klimata v vysokikh shirotakh i ee otrazhenie v kolebaniyakh klimata i lednikov okruzhayushchikh raionov (Structure of climatic variations at higher latitudes and its reflection in the variation of climate and glaciers of surrounding regions). *Trudy GGO*, vol. 378, p. 3–16.

17. Drozdov, O.A. 1977. K voprosu o real'nosti sverkhvekovykh trendov v meteorologicheskikh ryadakh i ikh proiskhozhdeniye (The reality of supersecular trends in meteorological series and their origin). *Trudy GGI*, vol. 247, p. 3–8.

18. Borzenkova, I.I. et al. 1976. Izmenenie temperatury vozdukha severnogo polushariya za period 1881–75 (Change in the air temperature of the northern hemisphere for the period 1881–75). *Meteorologiya and Gidrologiya*, No. 7, p. 27–35.

19. Kupriyanov, V.V. (Ed.). 1978. Izuchenie snezhnogo pokrova po dannym sputnikovoi informatsii (A study of ice cover from data of satellite

information). *Trudy GGI*, vol. 243, p. 1–79.

20. Andrianov, Yu.G. et al. 1973. Infrakrasnye spektry izlucheniya Zemli v kosmos (Infrared spectrum of earth's radiation in space). Sovetskoe Radio, Moscow. 112 p.

21. Kupriyanov, V.V. (Ed.). 1976. Ispol'zovanie sputnikovoi informatsii v gidrologii (Application of satellite information in hydrology). *Trudy GGI*, vol. 237, p. 1–196.

22. Kalinin, V.P., Yu.V. Kurilova and P.A. Kolosov. 1977. Kosmicheskie metody v gidrologii (Spatial methods in hydrology). Gidrometeoizdat, Leningrad. 184 p.

23. Kondrat'ev, K.Ya. 1968. Meteorologicheskie issledovaniya (Meteorological investigations). In *Uspekhi SSSR v issled. kosmich. prostranstva.* Nauka, Moscow. p. 47–70.

24. Kondrat'ev, K.Ya. and L.N. D'yachenko. 1970. Sravnenie eksperimental'nykh u paschetnykh velichin ukhodyashchego dlinnovolnovogo izlucheniya dlya razlichnykh sezonov goda (A comparison of the experimental and computed values of outgoing long-wave radiation in different seasons of the year). *Trudy GGO*, vol. 252, p. 35–43.

25. Kondrat'ev, K.Ya., L.N. D'yachenko and K.Ya. Vinnikov. 1970. Klimatologiya radiatsionnogo balansa Zemli na sovremennom etape (Modern climatology of earth's radiation balance). *Problemy fiziki atmosphery*, vol. 7, p. 21–61.

26. Kondrat'ev, K.Ya. and G.A. Nikol'skii. 1970. Variatsii solnechnoi postoyannoi po aerostatnym issledovaniyam v 1962–68 gg (Study of variation of solar constant from balloon investigations during years 1962–68). *Izv. AN SSSR. fizika atmosphery i okeana*, vol. 6, No. 3, p. 227–238.

27. Kondrat'ev, K.Ya. 1971. Sputnikovaya meteorologiya (Satellite meteorology). In *Meteorologiya i Klimatologiya.* Moscow. vol. 1, p. 25–85.

28. Kondrat'ev, K.Ya. 1971. Sputnikovaya klimatologiya (Satellite climatology). Gidrometeoizdat, Leningrad. 64 p.

29. Kondrat'ev, K.Ya. 1971. Solnechnaya postoyannaya (Solar constant). *Meteorologiya u Gidrologiya*, No. 3, p. 8–14.

30. Kondrat'ev, K.Ya. and G.A. Nikol'skii. 1973. Solnechnaya postoyannaya i ee vozmozhnye variatsii (Solar constant and its probable variations). In *Trudy simpoz. po solnechnokorpuskulyarn. e'ffektam v troposphere i stratosphere na XV General Assamblee MGGS*, 1971, Gidrometeoizdat, Leningrad, p. 143–148.

31. Kondrat'ev, K.Ya. and G.A. Nikol'skii. 1979. Solnechnaya postoyannaya i klimat (Solar constant and climate). *DAN SSSR*, vol. 243, No. 3, p. 607–610.

32. Kondrat'ev, K.Ya. 1974. Sputnikovaya meteorologiya (Satellite meteo-

rology) (1970–1972 gg). In *Meteorologiya i Klimatologiya*. Moscow. vol. 2, p. 32–190.

33. Kondrat'ev, K.Ya. 1977. Sovremennye izmeneniya klimata i opredely-ayushchiye ikh factory (Contemporary changes in climate and factors determining them). *Itogi Nauki u Tekhniki. Meteorologiya u Klimatologiya*, VINITI, Moscow. vol. 4, 202 p.

34. Kondrat'ev, K.Ya. 1978. Sputnikovyi monitorii klimata (Satellite monitoring of climate) Obninsk. VNIIGMI-MTsD. 52 p.

35. Obukhov, A.M. and V.M. Kovtunenko (Ed.). 1974. Kosmicheskaya strela. Opticheskie issledovaniya atmosfery (Space indicator. Optical investigations of atmosphere). Nauka, Moscow. 327 p.

36. Loshchilov, V.S. 1977. Ispol'zovanie microvolnovykh sputnikovykh izmerenii dlya kartirovaniya morskikh l'dov (The application of microwave satellite measurements on mapping of sea-ice). *Trudy AANII*, vol. 343, p. 40–45.

37. Le'm, Kh. Kh. 1976. Dlinnoperiodichnye kolebaniya klimata (Long-term variations of climate). *Byul. VMO*, vol. 25, No. 1, p. 3–10.

38. Lyubarskii, A.N. 1977. Kolebaniya ledovitosti severnykh morei i ikh vozmozhnye fizicheskie prichiny (Variations in ice content of Northern Sea and its probable physical reasons). *Trudy GGO*, vol. 386, p. 111–121.

39. Makarova, E.A. and A.V. Kharitonov. 1972. Raspredelenie energii v spektre Solntsai solnechnaya postoyannaya (Distribution of energy in the solar spectrum and solar constant). Nauka, Moscow. 288 p.

40. Malkevich, M.S. 1973. Opticheskie zondirovaniya atmosfery so sputnikov (Optical sounding of atmosphere from satellites). Nauka, Moscow. 303 p.

41. Markina, N.G., L.A. Pakhomov and L.A. Pakhomova. 1977. Nekotorye kharakteristiki ukhodyashchego izluchenia po izmereniyam so sputnika (Some characteristics of outgoing radiation obtained from measurements from satellites). *Trudy GosNITsIPR*, vol. 3, p. 95–104.

42. Myagchenkova, O.G. and V.I. Tulupov. 1971. Ukhodyashchee izluchenie Zemli v intervale dlin voln 15–28 μm po izmereniyam so sputnika "Kosmos" (Outward emission from the earth in the wavelength range of 15–28 μm as measured from the sputnik 'Kosmos'). *Geomagnetizm i ae'ronomiya*, vol. 11, No. 3, p. 401–405.

43. Berlyand, T.G. et al. 1977. Novoe v izuchenii padiatsionnogo rezhima i dinamiki klimata (New achievements in the study of the radiation regime and dynamics of climate). In *Sovremen. fundament. i prikl. issled. Gl. geofiz. obs Gidrometeoizdat*. Leningrad. p. 126–133.

44. Kondrat'ev, K.Ya. et al. 1973. O vozmozhnosti opredeleniya al'bedo podstilayushchei poverkhnosti po dannym sputnikovykh nablyudenii (The possibilities of determining the albedo of underlying surface from satellite observations). *Trudy GGO*, vol. 295, p. 62–78.

45. Ovchinnikov, V.I. 1978. O vychislenii statisticheskikh kharakteristik polya ukhodyashchego teplovogo izlucheniya (Computation of statistical characteristics of outgoing thermal radiation field). *Trudy GosNITsIPR*, vol. 9, p. 52–60.

46. Pivovarova, Z.I. 1975. Vekovoi khod pryamoi solnechnoi radiatsii (Secular pace of solar radiation). *Trudy GGO*, vol. 338, p. 39–60.

47. Pivovarova, Z.I. 1977. Radiatsionnye kharakteristiki klimata SSSR (Radiation characteristics of climate of USSR). Gidrometeoizdat, Leningrad. 335 p.

48. Chizhov, A.H. et al. 1977. Primenenie radiolokatsionnogo metoda pri izmerenii tolshchiny ledyanogo pokrova rek, ozer i vodokhranilishch (Application of radiolocation method to determine thickness of ice-cover on rivers, lakes and water reservoirs). *Trudy GGI*, vol. 245, p. 3–29.

49. Provorkin, A.V. 1977. Ispol'zovaniye snimkov, poluchennykh s meteoro-logicheskikh sputnikov, v kachestve osnovy dlya sostavleniya ledovykh kart (Use of pictures obtained from meteorological satellites as base in compilation of ice maps). *Trudy AANII*, vol. 343, p. 34–39.

50. Rubinshtein, E.S. 1973. Struktura kolebanii temperatury vozdukha na severnom polusharii (Structure of air temperature variations in northern hemisphere). Gidrometeoizdat, Leningrad. 34 p.

51. Rubinshtein, E.S. 1977. Nekotorye zamechaniya k stat'e I.I. Borzenko-voi, K.Ya. Vinnikova, L.P. Spirinoi, D.I. Stekhnovskogo 'Izmenenie temperatury vozdukha severnogo polushariya za period 1881–1975gg' (Some observations on the paper by I.I. Borzenko et al. "Changes in air temperature of northern hemisphere during 1881–1975"). *Meteorologiya i Gidrologiya*, No. 9, p. 106–110.

52. Rubinshtein, E.S. 1977. Struktura kolebanii temperatury vozdukha na severnom polusharii (The structure of air temperature variations in northern hemisphere). Gidrometeoizdat, Leningrad. p. 26.

53. Severova, V.A. and A.D. Chistyakov. 1977. Izmenilsya li temperaturnyi rezhim Moskvy za poslednie 100 let? (Has the temperature regime of Moscow undergone changes during the last 100 years?). *Chelovek i Stikhiya*, p. 49–50.

54. Sivkov, S.I. 1967. Metody rascheta kharakteristik solnechnoi radiatsii (Methods for computing solar radiation characteristics). Gidrometeo-izdat, Leningrad. 232 p.

55. Kondrat'ev, K.Ya. et al. (Ed.). 1975. Sovetskoamerikanskii eksperiment "Bering" (Soviet-American experiment "Bering"). Gidrometeoizdat, Leningrad. 315 p.

56. Budyko, M.I. et al. 1978. Teplovoi balans Zemli (Earth's thermal budget). Gidrometeoizdat, Leningrad. 41 p.

57. Flon, G. 1977. Istoriya i intranzitivnost' klimata (History and intransi-tiveness of climate). In *Fizicheskie osnovy klimata i ego modelirovaniya*.

Gidrometeoizdat, Leningrad. p. 114–124.

58. Agee, E.M. 1979. Present climatic cooling and its most probable cause. *Preprint. Symp. Empirical and Model Assisted Diagnosis of Climate and Climate Change*, Oct. 15–19, Tbilisi. 32 p.

59. Alexander, R.C. and R.L. Mobley, 1976. Monthly average sea-surface temperatures and ice-pack limits on a 1° global grid. *Mon. Wea. Rev.*, vol. 104, No. 2, p. 143–148.

60. Angell, J.K. and J. Korshover. 1978. Global temperature variation, surface to 100 mb: an update into 1977. *Mon. Wea. Rev.*, vol. 106, No. 6, p. 755–770.

61. Angell, J.K. and J. Korshover. 1976. Variation of sunshine duration over the contiguous United States between 1950 and 1972, *J. Appl. Meteorol.*, vol. 14, No. 6, p. 1174–1181.

62. Angell, J.K. and J. Korshover. 1976. Global analysis of recent total ozone fluctuations. *Mon. Wea. Rev.*, vol. 104, No. 1, p. 63–75.

63. Angell, J.K. and J. Korshover. 1977. Estimate of the global change in temperature, surface to 100 mb between 1958 and 1975. *Mon. Wea. Rev.*, vol. 105, No. 4, p. 375–385.

64. Arking, A. 1978. The relative information content of wide, medium and narrow angle measurements of earth radiation energy parameters from satellite altitudes. *Third Conf. Atmos. Rad.*, June 28–30, Davis, Calif.

65. Barrett, E.C. 1974. Climatology from satellites. London: Methuen. 418 p.

66. Baumgartner, A., H. Meyer and W. Metz. 1976. Globale Verteilung der Oberflachenalbedo. *Meteorol. Rund*, H. 2, S. 38–43.

67. Bischoff, W. 1977. Comparability of CO_2, measurements. *Tellus*, vol. 29, No. 5, p. 435–444.

68. Blamont, J.E., J.P. Pommereau and G. Souchon. 1975. Ultraviolet and visible solar flux measurements on board 'Concorde.'—Final Rep. Dept. Transportation Contr. N00014 73 c 153. *Verrières le Buisson*, June, 62 p.

69. Bogus, K. 1975. Solar constant, AMO spectral irradiance and solar-cell calibration. *Techn. Memo. ESA TM-160 (ESTEC)*. European Space Agency. 9 p.

70. Bolle, H.J. 1973. Der Strahlungshaushalt der Atmosphäre nach Satelliten-messungen. *Ann. Meteorol. Neue Folge*, No. 6, p. 25–33.

71. Boville, B.W. 1978. Private communication.

72. Broeckner, W.S. 1975. Climatic change: are we on the brink of a pronounced climatic warming? *Science*, vol. 189, p. 460.

73. Bryson, R. 1974. A perspective on climate change. *Science*, vol. 184, p. 753–760.

74. Campbell, G.G. and T.H. von der Haar. 1975. Monthly average of global and zonal radiation budget from integrating sensors, an accuracy assessment from numerical simulation. *Proc. Second American Conf. on*

Radiation, Anaheim, Nov., p. 115–118.
75. Campbell, G.G. and T.H. von der Haar. 1978. Optimum satellite orbits for accurate measurements of the earth's radiation budget. Summary. *Colorado State Univ. Atmos. Sci. Pap.*, No. 289, 61 p.
76. Pittock, A.B. 1978. Climatic change and variability, a southern perspective. Cambridge Univ. Press, New York. 455 p.
77. Crommelynck, D. 1975. Approche objective d'une valeur de la constante solaire validée par détermination indirecte de la constante de Stefan-Boltzmann. *Publs. Inst. Roy. Meteorol., Belg.*, No. 91, p. 95–106.
78. Damon, P.E. and S.M. Kunen. 1976. Global cooling. *Science*, vol. 193, No. 4252, p. 447–453.
79. de Luisi, J.J. 1975. Measurements of the extraterrestrial solar radiant flux from 2981 to 4000 Å and its transmission through the earth's atmosphere as it is affected by dust and ozone. *J. Geophys. Res.*, vol. 80, No. 3, p. 345–354.
80. Kondrat'ev, K.Ya. et al. 1972. Determination of the albedo of the underlying surface from the data of albedo measurements from meteorological satellites. *Proc. Int., Radiation Symp.*, Tohoku Univ. p. 419–427.
81. Drummond, A.J. and M.P. Thekaekara (Eds.). 1973. The extraterrestrial solar spectrum. Mount Prospect, Illinois: Inst. Environ. Sci. 169 p.
82. Duhamel, T. and P. Marchal. 1979. Bilan radiatif de la Terre: simulation des flux et mesures accélerometriques. Rech. aérosp. No. 2, p. 119–127.
83. Ellis, J.S. and T.H. von der Haar. 1976. Zonal average earth radiation budget measurements from satellites for climate studies. *Atmos. Sci. Paper* N 240, Dep. Atmos. Sci.. Colorado State Univ., Fort Collins. 50 p.
84. Ellis, J.S. 1977. Cloudiness, the planetary radiation budget, and climate. Ph.D. Thesis, Dept. Atmos. Sci., Colorado State Univ., Fort Collins. 96 p.
85. Ellis, J.S. and T.H. von der Haar. 1978. Solar radiation reaching the ground determined from meteorological satellite data. *Third Conf. Atmos. Rad. Davis, Calif.*, June 28–30, p. 187–189.
86. Hickey, J.R. et al. 1976. Extraterrestrial solar irradiance measurements from the *Nimbus-6* satellite. In *Sharing the Sun*, Winnipeg. Aug. 5–20, 9 p.
87. Fabian, P. and Pruchniewicz. 1977. Meridional distribution of ozone in the troposphere and its seasonal variations. *J. Geophys. Res.*, vol. 82, No. 15, p. 2063–2073.
88. Falconer, P.D. and W. Peyinghaus. 1975. Radiative balance in the atmosphere as a function of season, latitude and height. *Arch. Meteorol. Geophys. und Bioklimatol.*, vol. 23, No. 3, p. 201–223.

89. Feddes, R.G. and K.N. Liou. 1977. Sensitivity of upwelling radiance in *Nimbus-6* HIRS channels to multilayered clouds. *J. Geophys. Res.* vol. 82, No. 37, p. 5977–5989.

90. Forgan, B.W. 1977. Solar constants and radiometric scales. *Applied Optics*, vol. 16, No. 6, p. 1628–1632.

91. Fröhlich, C. 1976. The solar constant: a critical review. *Preprint. Symp. on Radiation Garmisch-Partenkirchen*, Aug. 19–28, 5 p.

92. Fröhlich, C. 1977. Contemporary measures of the solar constant. In *The Solar Output and Its Variation/Ed.* O.R. White. Boulder, Colo. p. 93–109.

93. Green, R.N. and G.L. Smith. 1978. Deconvolution estimation theory applied to *Nimbus-6* ERB data. *Third Conf. Atmos. Rad.*, June 28–30, Davis, Calif. p. 376–379.

94. Gruber, A. 1973. Review of satellite measurements of albedo and outgoing long-wave radiation. *NOAA Techn. Memo. NES.*, No. 48, 12 p.

95. Guenther, B. 1978. Maintaining calibration of long-term satellite monitoring devices as performed on the ozone sensing BUV instruments. *4th Joint Conf. Sens. Environ. Pollutants*, New Orleans, La, 1977, Washington, D.C., p. 218–221.

96. Gupta, M.G. 1971. A comparative study of radiation data from *Nimbus-2. Indian J. Meteorol. and Geophys.*, vol. 22, No. 3, p. 467–468, 494.

97. Harley, W.S. and T. Jakobsson. 1976. Climatic trends in the Northern Hemisphere and their implications. Internal Report APRB41, No. 14, *Atmos. Environ. Service*, Canada. 38 p.

98. Harrison, E.F., G.G. Gibson and P. Minnis. 1978. Sampling analysis for the earth radiation budget satellite system mission based on orbital coverage and cloud variability. *Third Conf. Atmos. Rad.*, June 28–30, Davis, Calif. p. 353–356.

99. Heath, D.F. 1974. Recent advances in satellite observations of solar variability and global atmospheric ozone. *Proc. Int. Conf. on Structure, Composition and General Circulation of the Upper and Lower Atmospheres and Possible Anthropogenic Perturbations*, Jan. 14–25, 1974, Toronto. vol. 2, p. 1267–1292.

100. Heath, D.F. 1974. Recent advances in satellite observations of solar variability and global atmospheric ozone. *Preprint X-912-74-190 Goddard Space Flight Center*, Greenbelt, Maryland. 39 p.

101. Heroux, L. and J.E. Higgins. 1977. Summary of full-disc solar fluxes between 250 and 1940 Å. *J. Geophys. Res.*, vol. 82, No. 22, p. 3307–3311.

102. Hickey, J.R., F.J. Griffin and H.B. Howell. 1977. Two years of solar measurements from the *Nimbus-6* satellite. *Preprint. Proc. Int. Solar Energy Soc. Solar World Conf.*, Orlando, Florida. 6–9 June, 5 p.

103. Historical review of earth radiation budget studies and the scientific rationale leading to ERBSS. *Third Conf. Atmos. Rad.*, June 28–30, 1978, Davis, Calif.

104. Holopainen, E. 1975. Diagnostic studies on the interaction between the time-mean flow and the large-scale transient fluctuations in the atmosphere. Rep. No. 8, Dept. Meteorol., Univ. Helsinki, Helsinki. 15 p.

105. Holopainen, E.O. 1976. Energy balance of the earth. Geophys. Fluid Dynamics Lab. Paper No. 63, Monash University, Clayton, Victoria, Australia. 24 p.

106. Hoyt, D.V. 1975. The energy budget of the earth. *Coll. Abstr. Second Conf. on Atmospheric Radiation*, Arlington, Virginia. 29–31 Oct., p. 159–161.

107. Jensenius, J.S., J.J. Cahir and H.A. Panofsky. 1978. Estimation of outgoing long-wave radiation from meteorological variables accessible from numerical models. *Q.J. Roy. Meteorol. Soc.*, vol. 104, No. 439, p. 119–130.

108. Kaba, M. 1972. Sugárzási mezök kutotása meteorológiai müholdak mérési adatai alapján. *Időjárás*, vol. 76, No. 3, p. 160–174.

109. Kecter, D.R. and R.M. Pytkowicz. 1977. Natural and anthropogenic changes in the global carbon dioxide system. Global Chem. Cycles and Alterations by Man. *Rept. Dahlem Workshop*, Berlin. p. 99–120.

110. Kington, J.A. 1976. An introduction to an examination of monthly and seasonal extremes in the climatic record of England and Wales using historical daily synoptic weather maps from 1781 onward. *Weather*, vol. 31, No. 3, p. 72–78.

111. Kondrat'ev, K.Ya. 1969. Radiation in the atmosphere. New York: Academic Press. 912 p.

112. Kondrat'ev, K.Ya. and G.A. Nikolsky. 1970. Solar constant and solar activity. *Q.J. Roy. Meteorol. Soc.*, vol. 96, No. 409, p. 509–522.

113. Kondrat'ev, K.Ya. and L.N. Dyachenko. 1971. Modern stage of the development of radiation climatology of the earth. *Astronautical Res. Proc. XXIst IAF Congress*, North Holland Publ. Co., p. 1019–1026.

114. Kondrat'ev, K.Ya. 1972. Radiation process in the atmosphere. WMO Monogr. Geneva. No. 409, 214 p.

115. Kondrat'ev, K.Ya. and G.A. Nikolsky. 1975. Reproduction of IPS-1956 at the University of Leningrad. *Proc. Symp. Solar Radiat. Meas. and Instrum.*, Rockville, Md., 1973, vol. 1, p. 203–216.

116. Kukla, G.J. and H.J. Kukla. 1974. Increased surface albedo in the Northern Hemisphere. *Science*, vol. 183, No. 4126, p. 709–714.

117. Kukla, G.J. 1976. Global variation of snow and ice extent. *COSPAR Proc. Symp. Meteorological Observations from Space: their Contribution to the First GARP Global Experiment*, June 8–10, Philadelphia. p. 110–115.

118. Künzi, K.F. 1975. Measurements of snow cover over land with the *Nimbus-5* microwave spectrometer. *Preprint. Int. Symp. Remote Sensing*, Ann Arbor. 5 p.

119. Labs, D. and H. Neckel. 1968. The radiation of the solar photosphere from 2000 Å to 100 μm. Z. Astrophys., vol. 69, No. 1, p. 1–75.

120. Lal, M. and H.S. Rathor. 1972. On the evaluation of planetary albedo of the earth from turbidity measurements in India. Pure and Appl. Geophys., vol. 93, p. 191–196.

121. Lamb, H.H. 1975. Climatic changes and their possible impact on future planning. Build. Serv. Eng., vol. 42, p. 286–295; discuss. p. 295–298.

122. Lamb, H.H. Climate in the 1970s. Nature, vol. 259, No. 5544, p. 606.

123. Lamb, H.H. 1977. Climatic analysis. Phil. Trans. Roy. Soc. London, vol. B 280, No. 972, p. 341–350.

124. van Loon, H. and J. Williams. 1976. The connection between trends of mean temperature and circulation at the surface. Part II. Summer. Mon. Wea. Rev., vol. 104, No. 8, p. 1003–1011.

125. Lyall, I.T. 1977. Climatic trends of 1960–1976. J. Meteorol., vol. 2, No. 21, p. 260–261.

126. Madden, R.A. 1976. Estimates of the natural variability of time-averaged sea-level pressure. Mon. Wea. Rev., vol. 104, No. 7, p. 942–952.

127. Major, G. 1976. On the absorption of solar radiation by the earth's atmosphere. Contrib. Atmos. Phys., vol. 49, No. 3, p. 212–216.

128. Major, G. 1976. Atmospheric absorption of solar radiation as determined from satellite data. COSPAR Space Res. 16, Berlin, p. 43–46.

129. Major, G. 1977. A légkör sugarzásegyenlegének meghatározása müholdas mérésekböl. Idöjárás, vol. 81, No. 5, p. 288–292.

130. Mani, A. 1971. Radiation balance of the earth-atmosphere systems from satellite radiation measurements. Indian J. Meteorol. Hydrol. and Geophys., vol. 22, No. 3, p. 455–460, 493.

131. Mani, A., R.R. Kelkar and V. Srinivasan. 1975. Effects of clouds and particulates on infrared radiation fluxes in the atmosphere over India. Indian J. Meteorol., Hydrol. and Geophys., vol. 26, No. 2, p. 192–198.

132. Matson, M. 1977. Winter snow-cover maps of North America and Eurasia from satellite records. 1966–1976. NOAA Techn. Memo., NESS-84, Washington. D.C. 28 p.

133. McClain, E.P. 1975. Environmental research and applications using the very high resolution radiometer (VHRR) on the NOAA-2 satellite. A pilot project in Alaska. Climate Arct., Fairbanks, p. 415–429.

134. McGinnis Jr., D.F., J.A. Pritchard and D.R. Wiesnet. 1975. Snow depth and snow extent using VHRR data from the NOAA-2 satellite. NOAA Techn. Memo., NESS-63, Washington, D.C. 10 p.

135. McGinnis Jr., D.F. and S.R. Schneider. 1978. Monitoring spring ice break-up from space. Photogramm. Eng. and Remote Sensing, vol. 44, No. 1, p. 57–68.

136. von der Haar, T.H. et al., 1973. Measurements of solar energy reflected by the earth and atmosphere from meteorological satellites. Solar

Energy, vol. 14, No. 2, p. 175–184.

137. Gloersen, P. et al. 1975. Microwave maps of the polar ice of the earth, *Climate Arct.*, Fairbanks. p. 407–414.

138. Neckel, H. and D. Labs. 1975. Proposal for a consistent system of the basic solar radiation data. *Proc. Symp. on Solar Radiation, Measurements and Instrumentation*, Nov. 13–15, 1973, Smithsonian Inst. p. 326–349.

138a. Kukla, G.J. et al. 1977. New data on climatic trends. *Nature*, vol. 270, No. 5638, p. 573–580.

139. Smith, W.L. et al. 1977. *Nimbus-6* earth radiation budget experiment. *Appl. Optics*, vol. 16, No. 2, p. 306–318.

140. Jacobwitz, H. et al. 1978. On the determination of synoptic scale radiation fluxes from ERB narrow angle directional observations. *Third Conf. Atmos. Rad.*, 28–30 June, Davis, Calif. p. 162–163.

141. Oort, A.H. and T.H. von der Haar. 1976. On the observed annual cycle in the ocean-atmosphere heat balance over the Northern Hemisphere. *J. Phys. Oceanogr.*, vol. 6, No. 6, p. 781–800.

142. Painting, D.J. 1977. A study of some aspects of the climate of the Northern Hemisphere in recent years. *Meteorol. Office Sci. Paper*, No. 35, London. 25 p.

143. Pearman, G.I., R.J. Francey and P.J.B. Fraser. 1976. Climatic implications of stable carbon isotopes in free rings. *Nature*, vol. 260, No. 5554, p. 771–772.

144. Pearman, G.I. 1977. Further studies of the comparability of baseline atmospheric carbon dioxide measurements. *Tellus*, vol. 29, No. 2, p. 171–181.

145. Pearman, G.I. 1977. Measurement of atmospheric composition at the Australian baseline atmospheric monitoring station. *Proc. Clean Air Soc. Symp.*, Analytical Techniques in the Determination of Air Pollutants, Melbourne Univ. p. 16–22.

146. Pease, R.W. and D.A. Nickols. 1976. Energy balance maps from remotely sensed imagery. *Photogramm. Eng. and Remote Sensing*, vol. 42, No. 11, p. 1367–1373.

147. Péczely, G. 1974. Variability of monthly and annual mean temperatures on the earth. *Idöjárás*, vol. 78, No. 4, p. 202–209.

148. Peyinghaus, W. 1974. Eine numerische Berechnung der Strahlungsbilanz und die Strahlungserwärmung der Atmosphäre in meridional—vertikalschnitt-Bonner Meteorol. Abhandl. Bonn: *Ferd. Dümmlers Verlag*, No. 22, 64 p.

149. Pittock, A.B. 1974. Stratospheric temperature anomalies in 1963 and 1966. *Q.J. Roy. Meteorol. Soc.*, vol. 100, No. 423, p. 39–45.

150. Plass, G.N. and G.W. Kattawar. 1972. Effect of aerosol variation on radiance in the earth's atmosphere-ocean system. *Appl. Opt.*, vol. 11, No. 7, p. 1598–1604.

246

151. Raschke, E. and H.J. Preuss. 1979. The determination of the solar radiation budget at the earth's surface from satellite measurements. *Meteorol. Resch.*, vol. 32, No. 1, p. 21–28.
152. Raschke, E. 1979. The sun-earth observatory and climatology satellite (SEOCS)—results of a phase A study. *Proc. Int. Conf. Evolution of Planet. Atmos. and Climatol. of the Earth*, 16–20 Oct., 1978. CNES, Tolouse, France. p. 541–552.
153. Reiner, W.A. and H.E. Wright Jr. 1977. Impact of prehistoric and present fire patterns on the carbon dioxide content of the atmosphere. Global Chem. Cycles and Alterations by Man. *Rept. Dahlem Workshop*, Berlin. 1976, p. 121–135.
154. Duncan, CH.H. et al. 1977. Rocket calibration of the *Nimbus-6* solar constant measurements. *Appl. Optics.*, vol. 16, No. 10, p. 2690–2697.
155. Rockwood, A.A. and S.K. Cox. 1976. Satellite inferred surface albedo over Northwestern Africa. *Atmos. Sci. Paper No. 265*, Dept. Atmos. Sci., Colorado State Univ., Fort Collins. 64 p.
156. Sanderson, R.M. 1975. Changes in the area of Arctic sea ice 1966 to 1974. *Meteorol. Mag.*, vol. 104, No. 1240, p. 313–323.
157. Sasamori, T., J. London and D.V. Hoyt. 1974. Meteorology of the Southern Hemisphere. Radiation budget of the Southern Hemisphere. *Meteorol. Monogr.*, vol. 13, No. 35, p. 4–23.
158. Fischer, A.D. et al. 1976. Satellite observations of snow and ice with an imaging passive microwave spectrometer. *COSPAR Proc. Symp. on Meteorological Observations from Space: Their Contribution to the FIGGE*, Philadelphia. June 8–10, p. 98–103.
159. Satellite ozone measurements by the multi-filter radiometer (MFR) satellite system. *SOAC Bull.*, 1977, No. 4, p. 1–4.
160. Schaal, L.A. and R.F. Dale. 1977. Time of observation temperature bias and "climatic change". *J. Appl. Meteorol.*, vol. 16, No. 3, p. 215–222.
161. Schneider, S.H. and R.D. Dennett. 1975. Climatic barriers to long-term energy growth. *Ambio.*, vol. 4, p. 65–75.
162. Schultz, C. and W.L. Gates. 1974. Supplemental global climatic data: July. Rand Corp. 1700 Main St., Santa Monica, CA 90406, 38 p.
163. Scope Workshop on Biogeochemical Cycling of Carbon. March 21–26, 1977. vols. 1, 2, 3. Hamburg/Ratzeburg.
164. Siegenthaler, U. and H. Oeschger. 1978. Predicting future atmospheric carbon dioxide levels. *Science*, vol. 199, No. 4327, p. 388–394.
165. Smith, E.V.P. and D.M. Gotlieb. 1975. Solar flux and its variations. In *possible relationships between solar activity and meteorological phenomena*. NASA Washington, D.C. p. 97–118.
166. Künzi, K.F. et al. 1976. Snow and ice surfaces measured by the *Nimbus-5* microwave spectrometer, *J. Geophys. Res.*, vol. 81, No. 27, p. 4965–4976.

167. Sobczak, L.W. 1977. Ice movements in the Beaufort Sea 1973–1975; determination by ERTS imagery. *J. Geophys. Res.*, vol. 82, No. 9, p. 1413–1418.
168. Staelin, D.H. 1975. Passive microwave sensing of the earth. *IEEE-MTT-S Int. Microwave Symp. Microwaves Serv, Man.*, Palo Alto, Calif., New York. p. 20–22.
169. Stanford, J.L. 1974. Stratospheric water-vapor upper limits inferred from upper-air observations. Part I. Northern Hemisphere. *Bull. Amer. Meteorol. Soc.*, vol. 55, No. 3, p. 194–212.
170. Stanford, J.L. 1974. Possible long-term variations in stratospheric water-vapor content. *Weather*, No. 3, p. 107–112.
171. Starr, V.P. and A.H. Oort. 1973. Five-year climatic trend for the Northern Hemisphere. *Nature*, vol. 242, p. 310–313.
172. Streten, N.A. 1977. Seasonal climatic variability over the southern oceans. *Archiv Meteorol., Geophys., Bioklimat., Ser. B.*, vol. 25, No. 1, p. 1–20.
172a. Stuiver, M. 1978. Atmospheric carbon dioxide and carbon reservoir changes. *Science*, vol. 199, No. 4326, p. 253–258.
173. Suttles, J.T., L.M. Avis and P.G. Renfroe. 1978. Directional models for analysis of earth radiation budget measurements. *Third Conf. Atmos. Rad.*, June 28–30, Davis, Calif. p. 372–375.
174. Ellis, J.S. et al. 1978. The annual variation in the global heat balance of the earth. *J. Geophys. Res.*, vol. 83, No. C4, p. 1958–1962.
175. Smith, W.L. et al. 1974. The earth radiation budget (ERB) experiment. *Preprint. NOAA-NESS*, Washington, D.C. 23 p.
176. Smith, W.L. et al. 1977. The first year of planetary radiation budget measurements from the *Nimbus-6* experiments. *Preprint NOAA-NESS*, Washington, D.C. 8 p.
177. Jacobowitz, H. et al. 1978. The first eighteen months of planetary radiation budget measurements from the *Nimbus-6* ERB measurements. *Third Conf. Atmos. Rad.*, June 28–30, Davis, Calif. p. 164–166.
178. Thekaekara, M.P. 1975. The total and spectral solar irradiance and its possible variations. *Proc. Workshop: the Solar Radiation and the Earth's Atmosphere*, California Inst. Technol. 19–21 May, p. 232–263.
179. Thekaekara, M.P. 1976. Solar irradiance: total and spectral and its possible variations. *Appl. Optics.*, vol. 15, No. 4, p. 915–920.
180. Toward an internationally Coordinated Earth Radiation Budget Satellite Observing System: Scientific Uses and Systems Considerations. *COSPAR Report to ICSU and to JOC for GARP*, September 1978, 76 p.
181. Ulrich, R.K. 1975. Solar neutrinos and variations in the solar luminosity. *Proc. Workshop: the Solar Radiation and the Earth's Atmosphere*, California Inst. Technol., 19–21 May, p. 263–287.
182. Vernekar, A.D. 1972. Long-period global variations of incoming solar

248

radiation. *Meteorol. Monogr.* vol. 12, No. 34, p. 126.

183. von der Haar, T.H. 1973. Solar insolation microclimate determined using satellite data. *Solar Data Workshop, NOAA NSF-RA-N-72-062*, p. 143–147.

184. von der Haar, T.H. and A.H. Oort. 1973. New estimate of annual poleward energy transport by Northern Hemisphere oceans. *J. Phys. Oceanogr.*, vol. 3, No. 2, p. 169–172.

185. von der Haar, T.H. and E. Raschke. 1973. Measurements of the energy exchange between earth and space from satellites during the 1960s. *Ann. Meteorol. Neue Folge*, No. 6, p. 339–345.

186. von der Haar, T.H. 1975. Measurement of planetary albedo from satellites. *Proc. Symp. on Solar Radiation, Measurements and Instrumentation*, No. 13–15, 1973, Smithsonian Inst. p. 443–463.

187. von der Haar, T.H. and J. Ellis. 1975. Albedo of the cloud-free earth-atmosphere system. 2nd Conf. Atmos. Radiat., Arlington, Va., *Collect. Abstr.*, Boston, Mass. s.a. p. 107–110.

188. Webster, P.J. and J.L. Keller. 1975. Atmospheric variations: vacillations and index cycles. *J. Atmos. Sci.*, vol. 32, No. 7, p. 1283–1300.

189. Wiesnet, D.R. and M. Matson. 1975. Monthly winter snow line variation in the Northern Hemisphere from satellite records, 1966–75. *NOAA Techn. Memo. NESS-74*, Washington, D.C. 22 p.

190. Wiesnet, D.R. and M. Matson. 1976. A possible forecasting technique for winter snow cover in the Northern Hemisphere and Eurasia. *Mon. Wea. Rev.*, vol. 104, No. 7, p. 828–835.

191. Willson, R.C. 1978. Accurate solar constant determinations by cavity pyrheliometers. *J. Geophys. Res.*, vol. 83, No. C8, p. 4003–4008.

192. Winston, J.S. 1971. The annual course of zonal mean albedo as derived from ESSA 3 and 5 digitized picture data. *Mon. Wea. Rev.*, vol. 99, No. 11, p. 818–827.

193. Winston, J.S., W.L. Smith and H.M. Woolf. 1972. The global distribution of outgoing long-wave radiation derived from SIRS radiance measurements. *Second Conf. Atmos. Radiat.*, Fort Collins, Colorado, Prepar. vol. Boston, Mass. p. 221–227.

194. Winston, J.S. 1978. Earth-atmosphere radiative heating based on NOAA scanning radiometer measurements: review of a four year record. *Third Conf. Atmos. Rad., 28–30 June*, Davis, Calif. p. 158–161.

195. Winston, J.S. and A.F. Krueger. 1977. Diagnosis of the satellite-observed radiative heating in relation to the summer monsoon. *Pure and Appl. Geophys.*, vol. 115, No. 5–6, p. 1131–1144.

196. Woerner, C.V. 1978. The earth radiation budget satellite system: an overview. *Third Conf. Atmos. Rad.*, June 28–30, Davis. Calif. p. 345–348.

197. WMO Statement on Climatic Change. EC-XXVIII/PINK 33, Appendix B, Geneva, June 1976, p. 3–4.

198. Woodwell, G.M. 1978. The carbon dioxide question. *Sci. Amer.*, vol. 238, No. 1, p. 34–43.

199. Yates, H.W. 1977. Measurement of the earth radiation balance as an instrument design problem. *Appl. Optics*, vol. 16, No. 2, p. 297–299.

200. Zirin, H. and J. Walter (Eds.). May 1975. The solar constant and the earth's atmosphere. *Proc. of the Workshop*, Big Bear City, Calif. 332 p.

201. Zwally, H.J. and P. Gloersen. 1977. Passive microwave images of the polar regions and research applications. *Polar Record*, vol. 18, No. 116, p. 431–450.

CHAPTER II

1. Anolik, M.A. and A.M. Bunakova. 1972. Novyi variant rascheta intensivnosti sobstvennogo teplovogo izlucheniya sistemy planeta-atmosfera (sfericheskii sluchai) [A new variant of calculation of intensity of inherent thermal radiation of land-atmosphere system (spherical case)]. *Problemy fiziki atmosfery*, vol. 10, p. 79–86.

2. Gudi, R.M. 1966. Atmosfernaya radiatsiya (Atmospheric radiation). *Mir*, Moscow. 522 p.

3. Dmitriev, A.A. 1978. Vozmozhnye fizicheskie mekhanizmy geliogeofizicheskikh svyazei (Possible physical mechanisms of heliogeophysical relationships). *Trudy Gidromettsentra*, SSSR. vol. 37, p. 86–92.

4. Zuev, V.E. 1970. Rasprostranenie vidimykh i infrakrasnykh voln v atmosfere (Propagation of visible and infrared waves in atmosphere). *Sovetskoe radio*, Moscow. 496 p.

5. Carol, I.L. 1974. Vysotnye samolety i stratosfera (High altitude aeroplanes and stratosphere). Gidrometeoizdat, Leningrad. 49 p.

6. Carol', I.L. and A.A. Kiselev. 1977. Chislennaya model' global'nogo perenosa freonov v atmosfere i otsenki ikh vliyaniya na stratosfernyi ozon (A numerical model of global transition of Freon in the atmosphere and estimation of its effect on stratospheric ozone). *Trudy GGO*, vol. 394, p. 44–55.

7. Budyko, M.I. 1974. Klimat i vozdeistviya na aérozol'nyi sloi stratosfery (Climate and effect on aerosol layer of stratosphere). Gidrometeoizdat, Leningrad. 42 p.

8. Kondrat'ev, K.Ya. 1956. Luchistyi teploobmen v atmosfere (Radiant heat exchange in atmosphere). Gidrometeoizdat, Leningrad. 420 p.

9. Kondrat'ev, K.Ya. and K.E. Yakushevskaya. 1963. K voprosu o spectral'nom raspredelenii ukhodyashchego izlucheniya (Spectral distribution of outgoing radiation). *Problemy fiziki atmosphery*, vol. 2, p. 48–67.

10. Kondrat'ev, K.Ya. 1965. Aktinometriya (Actinometry). Gidrometeoizdat, Leningrad. 691 p.

11. Kondrat'ev, K.Ya., D.V. Styro and V.F. Zhvalev. 1966. Luchistyi pri-

tok tepla v oblasti spectra 4–10 μm na razlichnykh urovnyakh v atmosfere (Radiant heat influx in the spectral range 4–10 μm at various levels of atmosphere). *Izv. AN SSSR Fizika atmosfery i okeana*, vol. 2, No. 2, p. 121–132.

12. Kondrat'ev, K.Ya., Kh.Yu. Niilisk and R.Yu. Noorma. 1968. O spectral'nom raspredelenii radiatsionnykh pritokov tepla v svobodnoi atmosfere (The spectral distribution of radiant heat influx in the free atmosphere). *Izv. AN SSSR. Fizika atmosfery i okeana*, vol. 4, No. 6, p. 599–608.

13. Kondrat'ev, K.Ya. and Yu.M. Timofeev. 1969. Chislennoe modelirovanie fynktsii spectral'nogo propuskaniya dlya uzkikh spectral'nykh intervalov 15 μm polosy CO_2 (Numerical modeling of functions of spectral transmission for narrow spectral interval of 15 μm CO_2 band). *Izv. AN SSSR, Fizika atmosfery i okeana*, vol. 5, No. 4, p. 377–384.

14. Kondrat'ev, K.Ya. and G.A. Nikol'skii. 1970. Variatsy solnechnoi postoyannoi po aérostatnym issledovaniyam v 1962–1968 gg (Study of variation of solar constant from balloon investigations during 1962–1968). *Izv. AN SSSR, Fizika atmosfery. okeana*, vol. 6, No. 3, p. 227–238.

15. Kondrat'ev, K.Ya. and G.A. Nikol'skii. 1973. Solnechnaya postoyannaya u ee vozmozhnye variatsii (Solar constant and its possible variations). In *Trudy simpoz po solnechnokorpuskulyarn. èffektam v troposfere u stratosfere na XV* General assamblee MGGS, 1971. Gidrometeoizdat, Leningrad. p. 143–148.

16. Kondrat'ev, K.Ya. and G.A. Nikol'skii. 1974. Vozmozhnye izmeneniya intensivnosti electromagnitnogo izlucheniya Solntsa (The possible variations in the intensity of solar electromagnetic radiations). In *Trudy Pervogo Vsesoyuz soveshch.* "Solnechno-atmosfernye svyazi v teorii klimata i prognozax pogody". Gidrometeoizdat, Leningrad. p. 128–135.

17. Kondrat'ev, K.Ya., N.I. Moskalenko and G.A. Nikol'skii. 1975. Sopostavlenie raschetnykh i èxperimental'nykh dannykh o vertikal'nom raspredelenii dlinnovolnovykh potokov radiatsii v dnevnoe vremya (Comparison of computed and experimental data on vertical distribution of long-wave radiation flux during daytime). *Problemy fiziki atmosfery*, p. 96–105.

18. Kondrat'ev, K.Ya. 1976. Aèrozol' i klimat (Aerosol and climate). *Trudy GGO*, vol. 381, p. 3–66.

19. Kondrat'ev, K.Ya. and N.I. Moskalenko. 1976. Analiz priblizhennogo metoda rascheta polei teplovogo izlucheniya planet (Analysis of approximate method of computation of thermal radiation fields of planets). *Trudy GGO*, vol. 363, p. 3–20.

20. Kondrat'ev, K.Ya. and D.V. Pozdnyakov. 1976. Stratosfera i freony (Stratosphere and Freon). *Izv. AN SSSR, Fizika atmosfery i okeana*, vol. 12, No. 7, p. 683–695.

21. Kondrat'ev, K.Ya. 1977. Sovremenniye izmeneniya klimata i opredely-ayushchie ikh faktory (Contemporary changes in climate and the determining factors). *Itogi nauki i tekhniki. Meteorologiya klimatologiya,* VINITI, Moscow. vol. 4, 202 p.

22. Kondrat'ev, K.Ya. and N.I. Moskalenko. 1977. Teplovoe izlucheniye planet (The thermal radiation of planets). Gidrometeoizdat, Leningrad. 263 p.

23. Kondrat'ev, K.Ya. and G.A. Nikol'skii. 1978. Solnechnaya aktivnost' i klimat (Solar activity and climate). *DAN SSSR,* vol. 243, No. 3, p. 607–610.

24. Mak-Iven, M. and L. Phillips. 1978. Khimiya atmosfery (Chemistry of atmosphere). *Mir,* Moscow. 375 p.

25. Kondrat'ev, K.Ya. and N.E. Ter-Markaryants. 1976. Polnyi radiat-sionnyi eksperiment (Complete radiation experiment). Gidrometeoizdat, Leningrad. 238 p.

26. Rakipova, L.R. 1977. Dinamika verkhnikh sloev atmosfery, vliyanie solnechnoi aktivnosti i antropogennykh faktorov na klimat (Dynamics of upper layers of atmosphere, effect of solar activity and anthropogenic factors on climate). In *Sovrem. fundament. i prikladn. issled. Gl. geofiz. obs.,* Gidrometeoizdat, Leningrad. p. 53–57.

27. Svatkov, N.M. 1974. Antropogennye faktory i izmenenie temperatury vozdukha u poverkhnosti Zemli (Anthropogenic factors and variation of air temperature in the vicinity of the earth's surface). *In Vliyanie mestn. prirod.-klimat. uslovii na proektir gorodov.* Gidrometeoizdat, Moscow, p. 79–80.

28. Smirnov, B.M. 1977. Vliyaniye cheloveka na atmosferu Zemli (Effects of human activities on the earth's atmosphere). *Priroda,* No. 4, p. 10–19.

29. Sochnev, V.G., V.F. Tulinov and S.G. Yakovlev. 1977. Formirovanie okisi azota v verkhnei atmosfere pod deistviem korpuskulyarnogo izlucheniya (Formation of nitrogen oxide in the outer atmosphere under the influence of corpuscular radiation). *Trudy nauchn.-issl. tsentra izuch. prirod, resursov,* vol. 3, p. 56–60.

30. Sutugin, A.G. and I.V. Petryanov. 1976. Khimicheskiye prevrashcheniya okislov ugleroda i azota v atmosfere (Chemical transformations of carbon dioxide and nitrogen in the atmosphere). In *Fizicheskie aspecty zagryazneniya atmosfery* Vil'nyus, Moklas. p. 28–33.

31. Uitten, R. and I. Poppov. 1977. Osnovy a'eronomii (Fundamentals of Aeronomy). Gidrometeoizdat, Leningrad. 407 p.

32. Feigel'son, E.M. 1970. Luchistyi teploobmen i oblaka (Radiant heat exchange and clouds). Gidrometeoizdat, Leningrad. 230 p.

33. Feigel'son, E.M. 1978. Potoki solnechnogo izlucheniya i oblaka (Fluxes of solar radiation and clouds). Gidrometeoizdat, Leningrad. 157 p.

34. Tal'roze, V.L. et al. 1978. Khimiko-kineticheskie kriterii vozdeistviya

na ozonosferu veshchestv estestvennogo i antropogennogo proiskhozh-deniya (Chemical-kinetic criterion of impact of matter of natural and anthropogenic origin on the ozone sphere). *Izv. AN SSSR, Fizika atmosfery i okeana*, vol. 14, No. 4, p. 355–365.

35. Ackerman, M., D. Frimont and C. Muller. 1977. Stratospheric me-thane—measurements and predictions. *Aeronomica Acta.*, A-N 180. Brussels, 24 p.

36. Friend, J.P. et al. 1975. Atmospheric effects. In *Long-term Worldwide Effects of Multiple Nuclear-Weapons Detonations*. Washington, D.C. NAS, p. 24–63.

37. Bach, W. 1976. Global air pollution and climatic change. *Revs. Geophys. Space Phys.* vol. 14, No. 3, p. 429–474.

38. Batten, E.S. 1966. The effects of nuclear war on the weather and climate. *The Rand Corp.*, RM-4989-*TAB*.

39. Bauer, E. and F.R. Gilmore. 1975. The effect of atmospheric nuclear explosions on total ozone. *Revs. Geophys. Space Phys.*, vol. 13, No. 4, p. 451–458.

40. Bauer, E. 1978. A catalog of perturbing influences on stratospheric ozone, 1955–1975. Inst. Defence analysis paper P-1340. Rep. N FAA-EQ-78-20 U.S. Dept. Transportation, Washington. D.C. 55 p.

41. Bernhardt, K.H. and F. Kortüm. 1976. Beenflüssung der Atmosphäre durch menschliche Aktivitäten. *Geod. Geoph. Veröff*, vol. 2, No. 21, p. 3–62.

42. Blake, A.J. and J.H. Carver. 1977. The evolutionary role of atmospheric ozone. *J. Atmos. Sci.*, vol. 34, No. 5, p. 720–728.

43. Boffey, P.M. 1975. Nuclear war: Federation disputes Academy on how bad effects would be. *Science*, vol. 190, p. 248–250.

44. Boughner, R.E. 1978. The effect of increased carbon-dioxide concentra-tions on stratospheric ozone. *J. Geophys. Res.*, 1978, vol. 83, No. C3, p. 1326–1333.

45. Boville, B.W. 1976. Stratospheric ozone layer research. *Proc. Fourth Conf. on CIAP*, Febr. 1975. Dept. Transportation, Washington, D.C. p. 55–57.

46. Broderick, A. 1977. Stratospheric effects from aviation. *AIAA SAE 13th Propulsion, Conf.*, Orlando, Florida, July 11–13. 13 p.

47. Butler, D.M. 1978/79. Input sensitivity study of a stratospheric photo-chemistry model. *Pure and Appl. Geophys*, vol. 117, No. 3, p. 430–435.

48. Butler, D.M. and R. Stolarski. 1978. Interaction of simultaneous per-turbations of stratospheric photochemistry. *Paper Presented WMO Symp. Geophys. Asp. and Conseq. Changes in the Compos. of the Stra-tosphere*, Toronto, 26–30 June 1978, WMO-N 511, p. 229–232.

49. Callis, L.B. 1978. On the coupled nature of atmospheric phenomena. *Paper Presented WMO Symp. Geophys. Asp. and Conseq. Changes in the*

Compos. of the Stratosphere, Toronto. 26–30 June 1978, WMO-N 511, p. 193–202.

50. Chameides, W.L. and R.J. Cicerone. 1978. Effects of non-methane hydrocarbons in the atmosphere. *J. Geophys. Res.*, vol. C83, No. 2, p. 947–952.

51. Chang, J.S. 1976. Uncertainties in the validation of parameterized transport in 1-D models of the stratosphere. *Proc. Fourth Conf. on CIAP*, Febr. 1975, Dept. Transportation, Washington, D.C. p. 175–182.

52. Chang, J.S., W.H. Duewer and D.J. Wuebbles. 1978. The atmospheric nuclear tests of the 50's and 60's: a significant test of ozone depletion theories. *Preprint UCRL-80246*, Lawrence Livermore Lab., Univ. Calif. September 1977. *J. Geophys. Res.*, Livermore, 34 p.

53. Chen, Ts.Ch. and V. Ramanathan. 1978. A numerical simulation of seasonal stratospheric climate. Part II Energetics. *J. Atmos. Sci.*, vol. 35, No. 4, p. 615–633.

54. Chow, M.D. and A. Arking. 1978. An infrared radiation routine for use in numerical atmospheric models. *Third Conf. Atmos. Radiation of the Amer. Meteorol Soc. Preprints.* June 28–30, Davis, Calif. p. 303–305.

55. Chow, S.H. and R.J. Curran. 1978. The effect of ground hydrology on climate sensitivity to solar constant variations. *Third Conf. Atmos. Radiation of the Amer. Meteorol. Soc. Preprints*, June 28–30, Davis, Calif. p. 335–338.

56. Christie, A.D. 1976. Atmospheric ozone depletion by nuclear weapons testing. *J. Geophys. Res.*, vol. 81, No. 15, p. 2583–2594.

57. CIAP Monographs. 1975. Washington, D.C. : Dept. Transportation, vols. 1-6.

58. Cicerone, R.J., R.S. Stolarski and S. Walters. 1974. Stratospheric ozone destruction by man-made chlorofluoromethanes. *Science*, vol. 185, No. 4157, p. 1165–1167.

59. Cicerone, R.J. 1975. Minor constituents in the stratosphere and mesosphere. *Revs. of Geophys. and Space Phys.*, vol. 13, No. 3, p. 900–924.

60. Coakley, J.A. and B.P. Briegleb. 1978. Accurate calculations of fluxes and cooling rates using emissivities. *Third Conf. Atmos. Radiation of the Amer. Meteorol. Soc. Preprints*, June 28–30, Davis, Calif. p. 179–181.

61. Coakley, Jr. J.A. 1978. Radiative-convective calculations of surface temperature changes caused by changes in stratospheric ozone. *Third Conf. Atmos. Radiation of the Amer. Meteorol. Soc. Preprints*, June 28–30, Davis, Calif. p. 289–292.

62. COVOS. 1976. Rapport Final. Activités 1972–1976. Presenté par Edmond A. Brun-Boulogne : Soc. Meteorol. France. 163 p.

63. Crutzen, P.J. 1974. Estimates of possible future ozone reductions from continued use of fluorochloromethanes (CF_2Cl_2, $CFCl_3$). *Geophys. Rev. Lett.*, vol. 1, No. 5, p. 205–208.

64. Crutzen, P.J., I.S.A. Isaksen and G.C. Reid. 1975. Solar proton events: stratospheric sources of nitric oxide. *Science,* vol. 189, p. 457–459.

65. Crutzen, P.J. 1976. Upper limits on atmospheric ozone reductions following increased application of fixed nitrogen to the soil. *Geophys. Rev. Lett.,* vol. 3, No. 3, p. 169–172.

66. Crutzen, P.J. and I.S.A. Isaksen. 1976. The impact of the chlorocarbon industry on the ozone layer. *Preprint. NCAR-NOAA,* Boulder, Colorado. 54 p.

67. Crutzen, P.J. 1976. A two-dimensional photochemical model of the atmosphere below 55 km: estimates of natural and man-made ozone perturbations due to NO_x. *Proc. 4th Conf. on CIAP,* Feb. 1975. Dept. Transportation, Washington, D.C. p. 264–279.

68. Derwent, R.G. and A.E.J. Eggleton. 1978. Ozone depletion estimates for global halocarbon and fertilizer usage employing one-dimensional modeling techniques. Harwell, Oxfordshire. 81 p.

69. Dickinson, R.E. 1977. Challenges in the atmospheric sciences. In *Space Shuttle Mission 80's.* Part I. San Diego, Calif. p. 463–470.

70. Duewer, W.H., D.J. Wuebbles and J.S. Chang. 1978. The effects of a massive pulse injection of NO_x into the stratosphere. *Preprint WMO Symp. on Geophys. Asp. and Conseq. Changes in the Compos. of the Stratosphere,* Toronto, Canada. June 26–30, 6 p.

71. Effects of chlorofluoromethanes on stratospheric ozone. *Assessment report,* NASA, Washington, D.C. Sept. 1977, 12 p.

72. Ellingson, R.G. and J.C. Gille. 1978. An infrared radiative transfer model, Part I. Model description and comparison of observations with calculations. *J. Atmos. Sci.,* vol. 35, No. 3, p. 523–545.

73. Ellsaesser, H.W. 1975. Sudden stratospheric warmings (SSW). *Preprint Lawrence Livermore Lab.,* Univ. Calif. June 13, 6 p.

74. Ellsaesser, H.W. 1976. Statement prepared for presentation at the 'Concorde' hearings before the Federal Aviation Administration. *Preprint Lawrence Livermore Lab.,* Univ. Calif. Jan. 5, 17 p.

75. Ellsaesser, H.W. 1976. Oral statement prepared at the 'Concorde' hearings before DOT secretary Coleman. *Preprint Lawrence Livermore Lab.,* Univ. Calif. Jan. 5, 4 p.

76. Ellsaesser, H.W. 1976. Ozone destruction by catalysis: credibility of the threat. *Preprint UCRL-78627. Lawrence Livermore Lab.* The Fall Annual Meeting of the Amer. Geophys. Union, San Francisco, California. Dec. 6–10, 14 p.

77. Ellsaesser, H.W. 1976. Ozone drop and depletion theories. *Preprint UCRL-77833. Lawrence Livermore Lab.,* Feb. 17, 2 p.

78. Ellsaesser, H.W. 1977. Has man increased stratospheric ozone? *Preprint UCRL-79544. Rev. I, Lawrence Livermore Lab.,* Aug. 10, 6 p.

79. Ellsaesser, H.W. 1978. A reassessment of stratospheric ozone: credibility

of the threat. *Climatic Change.*, vol. 1, No. 3, p. 257–266.

80. Ellsaesser, H.W. 1978. Comment on "Analysis of the independent variables in the perturbation of stratospheric ozone by nitrogen fertilizers" by Harold S. Johnston. *J. Geophys. Res.*, vol. 83, No. C4, p. 1983–1984.

81. Fels, S.B. and M.D. Schwarzkopf. 1978. Stratospheric effects of doubled CO_2 concentration in a general circulation model. *Third Conf. Atmos. Radiation of the Amer. Meteorol. Soc. Preprints*, June 28–30, Davis, Calif. p. 287–288.

82. Fels, S.B. 1978. An accurate, fast algorithm for calculating stratospheric CO_2 15 μm cooling rates in GCMS. *Third Conf. Atmos. Radiation of the Amer. Meteorol. Soc. Preprints*, June 28–30, Davis, Calif. p. 306–307.

83. Fluorocarbons and the environment. Report of Federal Task Force on Inadvertent Modification of the Stratosphere (IMOF) Council of Environ. Quality. Federal Council for Sci. and Technol. June 1975, 109 p.

84. Georgii, H.W. 1978. Highlights of the research in atmospheric chemistry during the last 25 years. *Pure and Applied Geophys.*, vol. 116, No. 2/3, p. 215.

85. Graham, R.A. and H.S. Johnston. 1977. The photochemistry of NO_3 and the kinetics of the N_2O_5–O_3 system. *Preprint. Dept. Chemistry and Materials and Molecular Research Division*, Univ. Calif., Berkeley. 45 p.

86. Wang, W.C. et al. 1976. Greenhouse effects due to man-made perturbations of trace gases. *Science*, vol. 194, No. 4226, p. 685–690.

87. Groves, K.S., S.R. Mattingly and A.F. Tuck. 1978. Increased atmospheric carbon dioxide and stratospheric ozone. *Nature*, vol. 273, p. 711–715.

88. Hampson, J. 1974. Photochemical war on the atmosphere. *Nature*, vol. 250, p. 189–191.

89. Harwood, R.S. and J.A. Pyle. 1977. Studies of the ozone budget using a zonal mean circulation model and linearized photochemistry. *Quart. J. Roy Meteorol. Soc.*, vol. 103, No. 436, p. 319–344.

90. Heath, D.F. and S.S. Prasad. 1976. Possible effects of solar sector structure on terrestrial stratospheric ozone field. *Proc. Int. Symp. Atmos.*, Ozone, Dresden, Berlin. 1977, vol. 2, 91 p.

91. Heath, D.F., A.J. Krueger and P.J. Crutzen. 1977. Solar proton event: influence on stratospheric ozone. *Science*, vol. 197, No. 4306, p. 886–889.

92. Heicklen, J. 1976. Atmospheric chemistry. New York: Academic Press, p. 406.

93. Hesstvedt, E. and I.S.A. Isaksen. 1978. Possible increase in total ozone due to anthropogenic release of carbon monoxide and nitrogen oxides near the ground. *Preprint Inst.*, Geophysics, Univ. Oslo. 7 p.

94. Hidalgo, H. 1976. Assessment of potential impact of stratospheric flight on earth's ultraviolet irradiance. *AIAAJ.*, vol. 14, No. 2, p. 137–149.

256

95. Hidalgo, H. and P.J. Crutzen. 1977. The tropospheric and stratospheric composition perturbed by NO_x emissions of high-altitude aircraft. *J. Geophys. Res.*, vol. 82, No. 37, p. 5833–5866.

96. Hidalgo, H. 1978. Understanding anthropogenic effects on ultraviolet radiation and climate. *IEEE Trans on Geosci. Electronics*, vol. GE-16, No. 1, p. 4–22.

97. High Altitude Pollution Program. A status report prepared in accordance with PL 95–96. *Rep. N FAA-AEQ-77-16*, U.S. Dept. Transportation, Washington, D.C. 1977, 18 p.

98. Hunten, D.M. 1976. The philosophy of one-dimensional modeling. *Proc. 4th Conf. on CIAP*, February 1975, Dept. Transportation, Washington, D.C. p. 147–155.

99. Johnston, H., G. Whitten and J. Birks. 1973. The effect of nuclear explosions on stratospheric nitric oxide and ozone. *Proc. 2nd Conf. on CIAP*, Washington, D.C. p. 340–350.

100. Johnston, H.S. 1975. Global ozone balance in the natural stratosphere. *Revs. Geophys. and Space Phys.*, vol. 13, No. 5, p. 637–649.

101. Johnston, H.S. 1975. Pollution of the stratosphere. *Annual Rev. of Physical Chemistry*, vol. 26, p. 315–338.

102. Johnston, H.S. and J. Podolske. 1978. Interpretation of stratospheric chemistry. *Revs. Geophys. and Space Phys.*, vol. 16, p. 491–524.

103. Johnston, H.S. and H. Nelson. 1977. Comment on "NO_x catalytic ozone destruction: sensitivity of rate coefficients" by W.H. Duewer, D.J. Wuebbles, H.W. Ellsaesser and J.S. Chang. *J. Geophys. Res.*, vol. 82, No. 18, p. 2593–2599.

104. Johnston, H.S. 1977. Expected short-term local effect of nuclear bombs on stratospheric ozone. *J. Geophys. Res.*, vol. 82, No. 21, p. 3119–3124.

105. Joseph, J.M. and R. Burtsztyn. 1976. A radiative cooling model in the thermal infrared for application to models of the general circulation. *J. Appl. Meteorol.*, vol. 15, No. 4, p. 319–325.

106. Kellogg, W.W. 1978. Effects of human activities on global climate. *WMO Techn. Note*, No. 156, 157 p.

107. Kellogg, W.W. and S.H. Schneider. 1978. Global air pollution and climate change. *IEEE Trans. on Geosci. Electronics*, vol. GE-16, No. 1, p. 44–50.

108. Kondrat'ev, K.Ya. 1969. Radiation in the atmosphere. New York Academic Press, 912 p.

109. Kondrat'ev, K.Ya. and G.A. Nikolsky. 1970. Solar constant and solar activity. *Quart. J. Roy. Meteorol. Soc.*, vol. 96, No. 409, p. 509–522.

110. Kondrat'ev, K.Ya. 1972. Radiation processes in the atmosphere, *WMO Monogr.* Geneva. No. 309, 214 p.

111. Kuo, H.L. 1978. The infrared cooling rate of the atmosphere. *Third*

Conf. Atmos. Radiation of the Amer. Meteorol. Soc. Preprints, June 28–30, Davis, Calif. p. 182–184.

112. Lacis, A.A. and J.E. Hansen. 1974. A parameterization for the absorption of solar radiation in the earth's atmosphere. *J. Atmos. Sci.*, vol. 31, No. 1, p. 118–133.

113. Lamb, H.H. 1977. *Climate: present, past and future. Volume 2, Climatic history and future.* London: Methuen; New York: Barnes and Noble. 835 p.

114. Landsberg, H.E. and E.S. Epstein. 1976. Concerning possible effects of air pollution on climate. *Bull. Amer. Meteorol. Soc.*, vol. 57, No. 2, p. 213–219.

115. Liu, S.C. et al. 1976. Limitation of fertilizer induced ozone reduction by the long lifetime of the reservoir of fixed nitrogen. *Geophys. Res. Lett.*, vol. 3, No. 3, p. 157–160.

116. Lovelock, J.E. and D.H. Pack. 1976. CCl_3F and CCl_4 data in the British Isles 1970–75. *Environmental Quarterly, Health and Safety Lab.*—302, US Energy Research and Development Administration, New York. April 1, p. 4–20.

117. Luther, F.M. and R.J. Gelinas. 1975. Effect of molecular multiple scattering and surface albedo on atmospheric photodissociation rates. *Preprint UCRL-75160, Lawrence Livermore Lab.*, Univ. Calif. 38 p.

118. McCracken, M.C. 1976. Climate-model results of stratospheric perturbations. *Proc. 4th Conf. on CIAP*, Feb., 1975, Dept. Transportation, Washington, D.C. 1976, p. 183–194.

119. Mahlman, J.D. 1976. Some fundamental limitations of simplified—transport models as implied by results from a three-dimensional general circulation/tracer model. *Proc. 4th Conf.* on *CIAP*, February 1975, Dept. Transportation, Washington, D.C., p. 132–146.

120. Mahlman, J.D., R.W. Sinclair and M.D. Schwarzkopf. 1978. Simulated response of the atmospheric circulation to a large ozone reduction. *Paper Presented WMO Symp. Geophys. Asp. and Conseq. Changes in the Compos. of the Stratosphere*, Toronto. 26–30 June, 1978, WMO-N 511. p. 219–220.

121. Manabe, S. and R.T. Wetherald. 1967. Thermal equilibrium of the atmosphere with a given distribution of relative humidity. *J. Atmos. Sci.*, vol. 24, No. 3, p. 241–259.

122. McElroy, M.B., S.C. Wofsy and Y.L. Yung. 1977. The nitrogen cycle perturbations due to man and their impact on atmospheric N_2O and O_3. *Philos. Trans. Roy. Soc. London B. Biol. Sci.*, vol. 277, No. 954, p. 159–181.

123. Measurement of the fundamental vibration-rotation spectrum of ClO. *Appl. Optics*, 1977, vol. 16, No. 3, p. 523–525.

124. Mintz, Y. and M. Schlesinger. 1976. Ozone production and transport

with the UCLA general circulation model. *Proc. 4th Conf. on CIAP*, February 1975. Dept. Transportation, Washington D.C. p. 201–223.

125. Mitchel, Jr. J.M. 1975. A reassessment of atmospheric pollution as a cause of long-term changes of global temperature. The Changing Global Environment. Ed. S.F. Singer. Dordrecht-Boston D. Reidel Publ. Co. p. 149–174.

126. Mitchel, Jr. J.M. 1976. An overview of climatic variability, and its causal mechanisms. *Quart. Res.*, vol. 6, No. 4, p. 1–13.

127. Modelès mathematiques de la stratosphere. M. Berlin. R. Borghi, G. Brasseur, R. Joatton, M. Maignan *COVOS Rep N6A. Etude de l'environment N25*, Aérospatiale Institut d'Aéronomie Spatiale de Belgique ONERA, Paris. 1976, 204 p.

128. Nicolet, M. 1976. On the production of nitric oxide by cosmic rays in the mesosphere and stratosphere. *Proc. 4th Conf. on CIAP*, Feb. 1975. Dept. Transportation, Washington, D.C. p. 292–302.

129. Goldsmith, P. et al. 1973. Nitrogen oxides, nuclear weapon testing, 'Concorde' and stratospheric ozone. *Nature*, vol. 244, p. 545–551.

130. Duewer, W.H. et al. 1977. NO_x catalytic ozone destruction: sensitivity to rate coefficients. *J. Geophys. Res.*, vol. 82, No. 6, p. 935–942.

131. Ohring, G. and Sh. Adler. 1978. Some experiments with a zonally averaged climate model. *J. Atmos. Sci.*, vol. 35, No. 2, p. 186–205.

132. McClatchey, R.A. et al. 1970. Optical properties of the atmosphere. *Environmental Res. Paper*, No. 331, AF Cambridge Res. Lab., Bedford, Mass. 85 p.

133. Callis, L.B. et al. 1977. Ozone: effect of UV variability and stratospheric coupling mechanism. *Proc. Joint Symp. Atmos. Ozone*, Dresden, 1976, vol. 2, Berlin. p. 331–332.

134. Paltridge, G.W. 1973. Direct measurement of water vapor absorption of solar radiation in the free atmosphere. *J. Atmos. Sci.*, vol. 30, No. 1, p. 156–160.

135. Parry, H.D. 1977. Ozone depletion by chlorofluoromethanes. Yet another look. *J. Appl. Meteorol.*, vol. 16, No. 11, p. 1137–1148.

136. Pratt, P.F. 1977. Effect of increased nitrogen fixation on stratospheric ozone. *Climatic Change*, vol. 1, No. 2, p. 109–136.

137. Ramanathan, V. 1975. Greenhouse effect due to chlorofluorocarbons: climatic implications. *Science*, vol. 190, p. 50–52.

138. Ramanathan, V., L.B. Callis and R.E. Boughner. 1976. Sensitivity of surface temperature and atmospheric temperature to perturbations in the stratospheric concentration of ozone and nitrogen dioxide. *J. Atmos. Sci.*, vol. 33, No. 6, p. 1092–1112.

139. Ramanathan, V. and W.L. Grose. 1977. A three-D circulation model study of the radiative-dynamic coupling within the stratosphere. *Beitr. Phys. Atmos.*, vol. 50, p. 55–70.

259

140. Ramanathan, V. and W.L. Grose. 1978. A numerical simulation of seasonal stratospheric climate. Part I. Zonal temperatures and winds. *J. Atmos. Sci.*, vol. 35, No. 4, p. 600–614.
141. Ramanathan, V. and R.E. Dickinson. 1979. The role of stratospheric ozone in the zonal and seasonal radiative energy balance of the earth-troposphere system. *J. Atmos. Sci.*, vol. 36, No. 6, p. 1084–1104.
142. Hampson, Jr. R.F. et al. 1978. Reaction Rate and Photochemical Data for Atmospheric Chemistry. 1977. *NBS Special Publ.*, Washington, D.C. May, No. 513, 106 p.
143. Reck, R.A. 1976. Stratospheric ozone effects on temperature, *Science*, vol. 192, No. 4239, p. 557–559.
144. Reck, R.A. 1978. Response of a radiative-convective temperature profile to variations in model physical parameters. Uncertainty in the calculated temperature from input data error. *Third Conf., Atmos. Radiation of the Amer. Meteorol. Soc. Preprints*, June 28–30, Davis, Calif. p. 339–340.
145. Reply to H.W. Ellsaesser's letter—"Ozone drop and depletion theories". *Preprint, NCAR*, Boulder, Colorado. March 4, 1976. 2 p.
146. Rodgers, C.D. 1975. Modeling of atmospheric radiation for climatic studies. The Physical Basis of Climate and Climate Modeling. *GARP Publ. Ser.*, No. 16, p. 177–180.
147. Rowland, Sh. 1975. Chlorofluoromethanes and stratospheric ozone—a scientific status report. *New Sci.*, vol. 68, No. 969, p. 8–11.
148. Rowland, F.S. and M.J. Molina. 1975. Chlorofluoromethanes in the environment. *Revs. Geophys. and Space Phys.*, vol. 13, No. 1, p. 1–35.
149. Ruderman, M.A. and J.W. Chamberlain. 1975. Origin of the sunspot modulation of ozone: its implications for stratospheric NO injection. *Planet and Space Sci.*, vol. 23, No. 2, p. 247–268.
150. Rundel, R.D., D.M. Butler and R.S. Stolarski. 1978. Uncertainty propagation in a stratospheric model. 1. Development of a concise stratospheric model. *J. Geophys. Res.*, vol. C83, No. 6, p. 3063–3074.
151. Schoeberl, M.R. and D.F. Strobel. 1978. The response of the zonally averaged circulation to stratospheric ozone reductions. *J. Atmos. Sci.*, vol. 35, No. 9, p. 1751–1757.
152. Schwarzkopf, M.D. and R.T. Wetherald. 1978. Sensitivity of a general circulation model to a change in shortwave radiation code. *Third Conf. Atmos. Radiation of the Amer. Meteorol. Soc. Preprints*, June 28–30, Davis, Calif. p. 284–286.
153. Scorer, R.S. 1977. The stability of stratospheric ozone and its importance. *Atmos. Environ.*, vol. 11, No. 3, p. 277–282.
154. Scorer, R.S. 1977. Halocarbons: environmental effect of chlorofluoromethane release. Report of the Committee on Impacts of Stratospheric Change. *Atmos. Environ.*, vol. 11, No. 7, p. 655–659.
155. Sheppard, P.A. 1958. The effect of pollution on radiation in the atmo-

sphere. *Int. J. Air Pollution*, vol. 1, p. 31–43.

156. Simon, P.C. 1978. Irradiation solar flux measurements between 120 and 400 μm. Current position and future needs. *Planet and Space Sci.*, vol. 26, No. 4, p. 355–366.

157. Singer, S.F. 1975. Pollution effects on global climate—an introduction. The Changing Global Environment/Ed. S.F. Singer, Dordrecht-Boston; D. Reidel Publ. Co. p. 7–12.

158. Smith, Jr. W.S. 1978. Uncertainties in evaluated atmospheric rate constants. *Paper Presented WMO Symp. Geophys. Asp. and Conseq. Changes in the Compos. of the Stratosphere*, Toronto, 26–30 June, 1978, WMO-N 511, p. 37–46.

159. McElroy, M.B. et al. 1976. Sources and sinks for atmospheric N_2O. *Revs. Geophys. and Space Phys.*, vol. 14, No. 2, p. 143–150.

160. Liu, S.C. et al. 1977. Sources and sinks of atmospheric N_2O and the possible ozone reduction due to industrial fixed nitrogen fertilizers— *Tellus*, vol. 29, No. 3, p. 151–163.

161. Stolarski, R.S. 1978. The impact of chlorofluoromethane and NO_x injections on stratospheric ozone. *Proc. 4th Joint Conf. on Sensing of Environ. Pollutants*, New Orleans, Nov. 6–11, 1977, Washington, D.C., p. 20–26.

162. Stolarski, R.S., D.M. Butler and R.D. Rundel. 1978. Uncertainty propagation in a stratospheric model. 2. Monte Carlo analysis of imprecisions due to reaction rates. *J. Geophys. Res.*, vol. C83, No. 6, p. 3074–3079.

163. Strobel, D.F. 1978. Photochemical-radiative damping and instability in the stratosphere. II. Numerical results. *Geophys. Res. Lett.*, vol. 5, No. 6, p. 523–525.

164. Sunde, J. and I.S.A. Isaksen. 1976. Stratospheric chlorine nitrate: reduced effect on the catalytic destruction of ozone due to molecular scattering and reflection of photon fluxes in the atmosphere. Rep. No. 21, Inst. Geofys. Univ. Oslo. 16 p.

165. Sundararaman, N. 1978. Uncertainties in the estimates of stratospheric impacts from aviation. *Pap. Presented WMO Symp. Geophys. Asp. and Conseq. Changes in the Compos. of the Stratosphere*, Toronto. 26–30 June, WMO. No. 511, 1978, p. 275–276.

166. Sze, N.D. and H. Rice. 1976. Nitrogen cycle factors contributing to N_2O production from fertilizers. *Geophys. Res. Lett.*, vol. 3, p. 343–346.

167. Sze, N.D. 1977. Anthropogenic CO emissions: implications for the atmospheric $CO-OH-CH_4$ cycle. *Science*, vol. 195, No. 4279, p. 673–675.

168. Theobald, A.G., R.G. Williams and M.J. Rycroft. 1977. The effects of changing fluxes of solar UV radiation and solar cosmic rays on the middle atmosphere temperatures and ozone concentration. In *Dyn. and Chem. Coupling between Neutral and Ionized Atmos.* Dordrecht-Boston. p. 79–84.

169. The possible impact of fluorocarbons and halocarbons on ozone. *Inter-dept. Comm. Atmos. Sci.*, Federal Council for Sci. and Technol. ICAS 18a-FY 75. May 1975. 75 p.

170. The Report of the Committee on Meteorological Effects of Stratospheric Aircraft (COMESA) 1972–1975, Part I, 256 p.; Part II, p. 257–597. London: Meteorol. Office, 1975.

171. Traugott, S.C. 1977. Infrared cooling rates for two-dimensional thermal perturbations in a non-uniform atmosphere. *J. Atmos. Sci.*, vol. 34, No. 6, p. 863–872.

172. Tuck, A.F. 1977. One-dimensional model studies of the effect of injected nitrogen oxides upon stratospheric ozone. *Proc. Joint Symp. Atmos. Ozone*, Dresden, 1976, Berlin. vol. 3, p. 65–86.

173. Tuck, A.F. 1978. Covariance and averaging of gases in 1, 2 and 3-dimensional atmospheric models. *Pap. Presented WMO Symp. Geophys. Asp. and Conseq. Changes in the Compos. of the Stratosphere*, Toronto, 26–30 June, 1978, WMO. No. 511, p. 247–254.

174. Valéry, N. 1976. The shape of war to come (SIPRI Yearbook Highlights Geophysical Warfare). *New Sci.*, vol. 70, No. 1005, p. 628–630.

175. Vupputuri, R.K.R. 1978/79. The structure of the natural stratosphere and the impact of chlorofluoromethanes on the ozone layer investigated in a 2-D time dependent model. *Pure and Appl. Geophys.*, vol. 117, No. 3, p. 448–485.

176. Wardle, D.I. and W.F.J. Evans. 1976. The effect of Freon usage on global climate: the greenhouse Freon effect. *Internal Rep. APRB 40 X 8*, *Atmos. Environ. Service*, Canada. 9 p.

177. Weill, G. and J. Christophe. 1977. Long-term variations of OH night-glow emission—relation to stratospheric humidity. In *Dyn. and Chem. Coupling between Neutral and Ionized Atmos.*, Dordrecht-Boston. p. 85–90.

178. Wilcox, R.W. 1978. Total ozone trend significance from space and time variability of daily Dobson data. *J. Appl. Meteorol.*, vol. 17, No. 3, p. 405–409.

179. Wofsy, S.C., M.B. McElroy and N.D. Sze. 1975. Freon consumption: implications for atmospheric ozone. *Science*, vol. 187, No. 4176, p. 535–537.

180. Wofsy, S.C., M.B. McElroy and Y.L. Yung. 1976. The chemistry of atmospheric bromine. *Proc. 4th Conf. on CIAP*, Feb. 1975, Dept. Transportation, Washington, D.C., p. 286–294.

181. Wu Man Li, C. 1976. Long-wave radiation and its effect on the atmosphere. A dissertation submitted to the faculty of the Div. Phys. Sci. Dept. Geophys. Sci. Univ. Chicago, Chicago. 160 p.

182. Wu Man Li, C., R. Godbole and L.D. Kaplan. 1978. Influence of systematic radiation differences on the dynamics of a model atmosphere. *Third*

262

Conf. Atmos. Radiation of the Amer. Meteorol. Soc. Preprints, June 28–30, Davis, Calif., p. 293–303.

CHAPTER III

1. Kondrat'ev, K.Ya. (Ed.). 1978. Atmosfernyi aérozol' i ego vliyanie na perenos izlucheniya (Atmospheric aerosol and its effect on radiative transfer). Gidrometeoizdat, Leningrad, 120 p.
2. Budyko, M.I. 1974. Izmeneniya klimata (Changes of climate). Gidrometeoizdat, Leningrad. 280 p.
3. Kondrat'ev, K.Ya., O.B. Vacil'ev and L.S. Ivlev. 1973. Vliyanie aerozolya na perenos izlucheniya: vozmozhnye klimaticheskie posledstviya (Effect of aerosol on radiative transfer: possible climatic consequences). Izd. LGU, Leningrad. 266 p.
4. Davitaya, F.F. 1965. O vozmozhnom vliyanii zapylennosti atmosfery na umen'shenie lednikov i potepleniye klimata (On possible effect of turbidity of atmosphere on the decrease of glaciers and warming of climate). Izv. AN SSSR. Ser. Geogr., No. 2, p. 3–22.
5. Kargin, B.A., S.V. Kuznetsov and G.A. Mikhailov. 1976. Otsenka coeffitsientov aerozol'nogo pogloshcheniya po izmereniyam pogloshchennoi radiatsii (Estimation of coefficients of aerosol absorption from the measurements of absorbed radiation). Fizika Atmosfery i Okeana, vol. 12, No. 7, p. 720–725.
6. Kozoderov, V.V. 1979. Effektivnye parametry atmosfernogo aerozolya v korotkovolnovoi oblasti spectra (Effective parameters of atmospheric aerosol in the shortwave region of spectrum). Trudy GosNITsIPR, vol. 5, p. 22–31.
7. Kondrat'ev, K.Ya. and O.I. Smoktii. 1973. O vliyanii aerozolya na spectral'noe al'bedo sistemy atmosfera-podstilayushchaya poverkhnost' (On the effect of aerosol on spectral albedo of atmosphere-underlying surface system). Izv. AN SSSR. Fizika Atmosfery i Okeana, vol. 9, No. 12, p. 1269–1282.
8. Kondrat'ev, K.Ya. and L.P. Rakipova. 1974. Radiatsiya i dinamika atmosfery: radiatsionnye effecty aerozolya (Radiation and atmosphere dynamics: radiation effects of aerosol). Trudy GGO, vol. 344, p. 64–82.
9. Kondrat'ev, K.Ya. 1977. Nekotorye aspekty fiziki sovremennykh izmenenii klimata (Some aspects of physics of contemporary climatic variations). Izv. AN SSSR. Fizika Atmosfery i Okeana, vol. 13, No. 3, p. 227–244.
10. Kondrat'ev, K.Ya. 1977. Sovremennye izmeneniya klimata i opredelyayushchie ikh faktory (Contemporary climatic changes and factors determining them). Itogi Nauki i Tekhniki. Meteorologiya I Klimatologiya, VINITI, Moscow. vol. 4, 202 p.

11. Livshits, G.Sh. 1977. Pogloshchenie atmosfernym aèrozolem v ul'tra-fioletovoi oblasti spectra (Absorption by atmospheric aerosol in the ultraviolet range of spectrum). *Izv. AN SSSR. Fizika Atmosfery i Okeana*, vol. 13, No. 11, p. 1214–1216.

12. Moskalenko, N.I. 1975. O vliyanii atmosfernogo aèrozolya na spektral'-noe i uglovoe raspredelenie teplovogo izlucheniya (On the effect of atmospheric aerosol on spectral and angular distribution of thermal radiation). *Izv. AN SSSR. Fizika Atmosfery i Okeana*, vol. 11, No. 12, p. 1254–1262.

13. Laktionov, A.G. et al. 1976. O svyazi opticheskikh i aerozol'nykh kharakteristik atmosfery vostochnoi chasti' ekvatorial'noi Atlantiki (On the relationship of optical and aerosol characteristics of atmosphere in the eastern part of the equatorial Atlantic). In *TROP'EKS-74*. Gidro-meteoizdat, Leningrad. vol. 1, p. 630–637.

14. Pivovarova, Z.I. 1977. Ispol'zovanie dannykh nazemnykh radiatsion-nykh nablyudenii dlya izucheniya prozrachnosti atmosfery (Application of data on ground radiation observations for the study of transparency of atmosphere). *Meteorologiya i Gidrologiya*, No. 9, p. 24–31.

15. Kondrat'ev, K.Ya. and H.E. Ter-Markaryants (Eds.). 1975. Polnyi radiatsionnyi eksperiment (Complete radiation experiment). Gidrome-teoizdat, Leningrad. 240 p.

16. Kondrat'ev, K.Ya. et al. 1977. Predvaritel'nye rezul'taty samoletnykh issledovanii po radiatsionnoi podprogramme AT'EP (Preliminary results of aircraft investigations on radiation subprogramme ATEP). *Meteoro-logiya i Gidrologiya*, No. 3, p. 97–104.

17. Snegirev, V.A. and V.V. Kozoderov, 1975. Raschet perenosa korotko volnovogo izlucheniya v atmosfere (Computation of shortwave radiative transfer in atmosphere). *Trudi Gidromettsentra SSSR*, vol. 165, p. 87–99.

18. Chapurskii, L.I. and A.P. Chernenko. 1977. Aèrozol'noe pogloshchenie izluchenii vidimogo i blizhnego IK diapazonov spectra v troposfere (Aerosol absorption of radiations of visible and middle infrared bands of spectrum in troposphere). *Trudy GGO*, vol. 393, p. 7–21.

19. Chapurskii, L.I. and A.P. Chernenko. 1977. Spektral'nye radiatsionnye kharakteristiki tropicheskoi atmosferi (Spectral radiation characteristics of tropical atmosphere). *Trudy GGO*, vol. 393, p. 22–30.

20. Ackerman, T.P., Kuo-Nan Liou and C.B. Leovy. 1976. Infrared radia-tive transfer in polluted atmospheres. *J. Appl. Meteorol.*, vol. 15, No. 1, p. 28–35.

20a. Aerosols and Climate. Report of the Meeting of the JOC Working Group. 7–11 Aug., 1978, Boulder, Colorado. Geneva: ICSU-WMO. Oct., 1978, 35 p.

21. Weiss, R. et al. 1977. Application of directly measured aerosol radiative properties to climate models. *In Proc. Symp. Radiation, Garmisch-Par-*

tenkirchen, August 19–28, 1976. Princeton: Sci. Press, p. 469–471.

22. Atwater, M.A. 1975. Climatic effects of spherical polydispersions. *Coll. Abstr. Second Conf. Atmospheric Radiation*, Arlington, Virginia. 29–31 Oct. p. 131–134.

23. Bednar, J. 1975. Model of radiation processes in atmospheric layers polluted by aerosol particles. *Stud. Geophys et Geod.* vol. 19, No. 2, p. 167–183.

24. Bensimon, J. and B. Dehove. 1977. Sensibilité de l'albedo planetaire a la pollution stratospherique en aérosol. Météorologie. No. 10, p. 15–20.

25. Bolin, B. and R.J. Charlson. 1976. On the role of the tropospheric sulfur cycle in the shortwave radiative climate of the earth. *Ambio.*, vol. 5, No. 2, p. 47–54.

26. Braslau, N. and J.V. Dave. 1973. Effect of aerosol on the transfer of solar energy through realistic model atmospheres. *J. Appl. Meteorol.*, vol. 12, No. 4, p. 601–619.

27. Braslau, N. and J.V. Dave. 1975. Atmospheric heating rates due to solar radiation for several aerosol-laden cloudy and cloud-free models. *J. Appl. Meteorol.*, vol. 14, No. 3, p. 396–399.

28. Cadle, R.D. and G.W. Grams. 1975. Stratospheric aerosol particles and their optical properties. *Rev. Geophys. Space Phys.*, vol. 13, No. 4, p. 475–501.

29. Callis, L.B. 1975. The earth albedo: the relative importance of stratospheric aerosols, ozone and water vapor. *Coll. Abstracts Second Conf. Atmospheric Radiation*, Arlington, Virginia. 29–31 Oct., p. 127–128.

30. Carlson, T.N. 1979. Atmospheric turbidity in Saharan dust outbreaks as determined by analyses of satellite brightness data. *Mon. Wea. Rev.*, vol. 107, No. 3, p. 322–335.

31. Charlson, R.J. et al. 1976. The dominance of tropospheric sulfate in modifying solar radiation. In *Atmospheric Aerosols: their Optical Properties and Effects*. A Topical Meeting on Atmospheric Aerosols. Williamsburg, Virginia. Dec. 13–15, MC6-1-5.

32. Chýlek, P. and J.A. Coakley, Jr. 1974. Aerosols and climate. *Science*, vol. 183, No. 4120, p. 75–77.

33. Chýlek, P. and J.A. Coakley, Jr. 1975. Man-made aerosols and the heating of the atmosphere over polar regions. In *Climate of the Arctic*, vol. 1. G. Weller, S.A. Bowling. Eds., Geophys. Inst. Univ. Alaska. p. 159–165.

34. Chýlek, P. and J.A. Coakley, Jr. 1975. Nonspherical aerosols and climate. *EOS*, vol. 56, No. 12, p. 997–1002.

35. Chýlek, P., G.W. Grams and R.G. Pinnick. 1976. Light scattering by irregular randomly oriented particles. *Science*, vol. 193, p. 480–482.

36. Coakley, Jr., J.A. and P. Chýlek. 1975. The two-stream approximation in radiative transfer: including the angle of the incident radiation.

J. Atmos. Sci., vol. 32, No. 2, p. 409–418.

37. Coakley, Jr. J.A. and G.W. Grams. 1975. Stratospheric aerosols and surface temperature. *Coll. Abstr. Second Conf. Atmospheric Radiation*, Arlington, Virginia, Oct. 29–31, p. 138–141.

38. Coakley, Jr. J.A. and G.W. Grams. 1976. Relative influence of visible and infrared optical properties of a stratospheric aerosol layer on the global climate. *J. Appl. Meteorol.*, vol. 15, No. 7, p. 679–691.

39. Dave, J.V. and N. Braslau. 1975. Effect of cloudiness on the transfer of solar energy through realistic model atmospheres. *J. Appl. Meteorol.*, vol. 14, No. 3, p. 388–395.

40. de Luisi, J.J. et al. 1976. Results of a comprehensive atmospheric aerosol-radiation experiment in the southwestern United States. Part I: Size distribution, extinction optical depth and vertical profiles of aerosols suspended in the atmosphere. Part II. Radiation flux measurements and theoretical interpretation/D.D. de Luisi, P.M. Furukawa, D.A. Gilette et al. *J. Appl. Meteorol.*, vol. 15, No. 5, p. 441–463.

41. Fischer, K. and H. Grassl. 1975. Absorption by airborne and deposited particles in the 8–13 micrometer range. *Tellus*, vol. 27, p. 522–528.

42. Galindo, I. 1977. Cambios climaticos antropogenicos en la cindad de Mexico y su repercusion en la salad publica: raquitismo por carencia. Memoria de la reunion sobre fluctuaciones climaticas y su impacto en las actividades humanas, Secondo Etapo, Mexico. p. 145–162.

43. Galindo, I. and A. Chávez. 1978. Thermal and radiational aspects of the Christman Eve 1977 air pollution episode in Mexico City. *Preprint. Inst. Geofis.*, UNAM, Mexico. 7 p.

44. Glazier, J., J.L. Monteith and M.H. Unsworth. 1975. Effects of aerosol on the local heat budget of the lower atmosphere. *Quart. J. Roy. Meteorol. Soc.*, vol. 101, No. 431, p. 95–102.

45. Gras, J.L. and J.E. Laby. 1979. Southern hemisphere stratospheric aerosol measurements. 2. Time variations and the 1974–1975 aerosol events. *J. Geophys. Res.*, vol. 84, N Cl, p. 303–307.

46. Grassl, H. 1975. Albedo reduction and radiative heating of clouds by absorbing aerosol particles. *Contr. Atmos. Phys.*, vol. 48, p. 199–210.

47. Halpern, P. and K.L. Coulson. 1976. A theoretical investigation of the effect of aerosol pollutants on shortwave flux divergence in the lower troposphere. *J. Appl. Meteorol.*, vol. 15, No. 5, p. 464–469.

48. Halpern, P. and K.L. Coulson. 1978. Solar radiation heating in the presence of aerosols. *IBM J. Res. and Develop.*, vol. 22, No. 2, p. 122–133.

49. Harshvardhan and R.D. Cess. 1976. Stratospheric aerosols: effect upon atmospheric temperature and global climate. *Tellus*, vol. 28, No. 1, p. 1–10.

50. Harshvardhan and R.D. Cess. 1977. Effect of tropospheric aerosols upon atmospheric infrared cooling rates. *Preprint. Laboratory for Planetary Atmospheres Research*, State Univ. of New York. Stony Brook, 22 p.
51. Herman, B.M. and S.R. Browning. 1975. The effect of aerosols on the earth-atmosphere albedo. *J. Atmos. Sci.*, vol. 32, No. 7, p. 1430–1445.
52. Henderson-Sellers, A. and A.J. Meadows. 1979. The zonal and global albedos of the earth. *Tellus*. vol. 31, No. 2, p. 170–173.
53. Hummel, J.R. 1977. Contribution to polar albedo of mesospheric aerosol layer. *J. Geophys. Res.*, vol. 82, No. 13, p. 1893–1900.
54. Idso, S.B. 1975. Low-level aerosol effects on earth's surface energy balance. *Tellus*, vol. 27, No. 3, p. 318–320.
55. Kondrat'ev, K.Ya. et al. 1978. Impact of aerosols on radiative energetics of tropical atmosphere *GARP Atlant. Trop. Exp. Proc. Int. Sci. Conf. Energ. Trop. Atmos.*, Tashkent, 1977. Geneva. p. 143–148.
56. Jennings, S.G., R.G. Pinnick and J.B. Gillespie. 1979. Relation between absorption coefficient and imaginary index of atmospheric aerosol constituents. *Appl. Opt.*, vol. 18, No. 10, p. 1368–1371.
57. Joseph, J.H. 1976. The effect of a desert aerosol on a model of the general circulation. *Preprint. Symp. Radiation*, Garmisch-Partenkirchen. Aug. 19–28, 5 p.
58. Kerschgens, M. 1978. Berechnungen des solaren Strahlungstransports in Atmosphäre und Ozean mit Hilfe ainer Zweistrommethode. Mitteilungen aus dem. Inst. Geophys. und Meteorol, der Univ. Köln, Köln. Jan. 1978, vol. 25, p. 103.
59. Kondrat'ev, K.Ya. 1972. Radiation processes in the atmosphere. Second IMO Lecture. WMO, No. 309, 214 p.
60. Kondrat'ev, K.Ya. et al. 1974. Spectral radiative flux divergence and its variability in the troposphere in the 0.4–2.4 μm region. *Appl. Optics*, vol. 13, No. 3, p. 478–486.
61. Kuhn, P.M., H.K. Weickmann and L.P. Stearns. 1975. Long-wave radiation effects of the Harmattan haze. *J. Geophys. Res.*, vol. 80, No. 24, p. 3419–3424.
62. Lenoble, J. 1977. La modification du bilan radiatif par les aérosols et poussières. Mec. fluides et environ. C.R. 14 èmes. *Journées hydraul.*, Paris, 1976. T.I. Paris. 1.12/1–1.12/6.
63. Lepel, E.A., K.M. Stefansson and W.H. Zoller. 1978. The enrichment of volatile elements in the atmospheres by volcanic activity: Augustine volcano 1976. *J. Geophys. Res.*, vol. C83, No. 12, p. 6213–6220.
64. Liou, K.N. and T. Sasamori. 1975. On the transfer of solar radiation in aerosol atmospheres. *J. Atmos. Sci.*, vol. 32, No. 11, p. 2166–2177.
65. Liou, K.N., K.P. Freeman and T. Sasamori. 1978. Cloud and aerosol effects on the solar heating rate of the atmosphere. *Tellus*, vol. 30, No. 1, p. 62–70.

66. Luther, F.M. 1974. Effect of stratospheric aerosol on solar heating rates. *Preprint. Lawrence Livermore Lab.*, Univ. Calif. June 18, 3 p.

67. Luther, F.M. 1974. Effect of increased stratospheric aerosol on planetary albedo. *Preprint. Lawrence Livermore Lab.*, Univ. Calif. UASG-74-14. June 27, 4 p.

68. Luther, F.M. 1974. Effect of stratospheric aerosol on long-wave radiative fluxes: variation with latitude and season. *Preprint. Lawrence Livermore Lab.*, Univ. Calif. UCID. 16559, July 19, 6 p.

69. Luther, F.M. 1975. Relative influence of stratospheric aerosols on solar and long-wave fluxes for a tropical atmosphere. *Preprint. UCLR-75760 (Rev. I). Lawrence Livermore Lab.*, Univ. Calif. July, 18 p.

70. Luther, F.M. 1975. Comparison of one-wavelength and average aerosol property calculations with full-flux solar radiation calculations. *Coll. Abstr. Second Conf. Atmospheric Radiation*, Arlington. Virginia. 29–31, Oct. p. 39–40.

71. Major, G. 1976. Effect of gases, aerosols and clouds on the atmospheric absorption of solar radiation. *Beitr. Phys. Atmos.*, vol. 49, No. 3, p. 216–221.

72. Major, G. 1976. The role played by the atmospheric aerosol in the warming of the troposphere. Idójárás. vol. 80, No. 5, p. 274–278.

73. Mugnai, A., G. Fiocco and G. Grams. 1978. Effects of aerosol optical properties and size distribution on heating rates induced by stratospheric aerosols. *Quart. J. Roy. Meteorol. Soc.*, vol. 104, p. 783–796.

74. Oliver, R.C. 1976. On the response of hemispheric mean temperature to stratospheric dust: an empirical approach. *J. Appl. Meteorol.*, vol. 15, No. 9, p. 933–950.

75. Paltridge, G.W. and C.M.R. Platt. 1976. Radiative processes in meteorology and climatology. Amsterdam-Oxford-New York: Elsevier Publ. Sci. Co. 318 p.

76. Prospero, J.M. 1979. Mineral and sea salt aerosol concentrations in various ocean regions. *J. Geophys. Res.*, vol. 84, No. C2, p. 726–732.

77. Pueschel, R.F. and P.M. Kuhn. 1975. Infrared absorption of tropospheric aerosols: urban-rural aerosols of Phoenix, Arizona. *J. Geophys. Res.*, vol. 80, No. 21, p. 2960–2963.

78. Raschke, E. 1975. Theoretical investigations of the absorption of solar radiative energy in the atmosphere-ocean system. *Proc. Symp. on Solar Radiation. Measurements and Instrumentation*, Nov. 13–15, 1973, Smithsonian Inst., p. 350–360.

79. Reck, R.A. 1974. Influence of surface albedo on the change in the atmospheric radiation balance due to aerosols. *Atmos. Environ.*, vol. 8, No. 8, p. 823–833.

80. Reck, R.A. 1975. Aerosols and polar temperature change. *Science*, vol. 188, p. 728–730.

81. Reck, R.A. 1975. Influence of aerosol cloud height on the change in the atmospheric radiation balance due to aerosols. *Atmos. Environ.*, vol. 9, No. 1, p. 89–99.

82. Reck, R.A. 1976. Thermal and radiative effects of atmospheric aerosols in the Northern Hemisphere calculated using a radiative-convective model. *Atmos. Environ.*, vol. 10, No. 8, p. 611–618.

83. Reynolds, D.W., T.H. von der Haar and S.K. Cox. 1975. The effect of solar radiation absorption in the tropical atmosphere. *J. Appl. Meteorol.*, vol. 14, No. 4, p. 433–444.

84. Russel, P.B. and G.W. Grams. 1975. Application of soil dust optical properties in analytical models of climate change. *J. Appl. Meteorol.*, vol. 14, No. 6, p. 1037–1043.

85. Sekera, Z. and L.L. Stowe, Jr. 1973. Effects of particulate matter on the radiance of terrestrial infrared radiation theory. *Beitr. Phys. Atmos.*, vol. 46, No. 2, p. 101–111.

86. Shaw, G.E. 1975. Radiative properties of atmospheric aerosols at Mauna Loa observatory. *Rep. UAG R-238*, Univ. Alaska, Fairbanks, Alaska. 55 p.

87. Shettle, E.P. 1976. Radiative transfer in rural, urban, and maritime model atmospheres. Atmospheric Aerosols: their Optical Properties and Effects. A Topical Meeting on Atmospheric Aerosols. Williamsburg, Virginia. Dec. 13–15, WA15-1-4.

88. Reagan, J.A. et al. 1978. Some results of the UA-ARE program. *Third Conf. Atmos. Rad.*, June 28–30, 1978, Davis, Calif. Boston. p. 241–243.

89. Lenoble, J. (Ed.). 1977. Standard Procedures to Compute Atmospheric Radiative Transfer in a Scattering Atmosphere. Boulder, Colo. 125 p.

90. Stephens, G.L. 1976. An improved estimate of the IR cooling in the atmospheric window region. *J. Atmos. Sci.*, vol. 33, No. 5, p. 806–809.

91. Stowe, L.L. 1974. Effects of particulate matter on the radiance of terrestrial infrared radiation results. *J. Atmos. Sci.*, vol. 31, No. 3, p. 755–767.

92. Pollack, J.B. et al. 1976. Stratospheric aerosols and climatic change. *Nature*, vol. 263, No. 5578, p. 551–555.

93. Turko, R.P. et al. 1979. A one-dimensional model describing aerosol formation and evolution in the stratosphere. I. Physical processes and mathematical analogs, *J. Atmos. Sci.*, vol. 36, No. 4, p. 699–717.

94. Toon, O.B. et al. 1979. A one-dimensional model describing aerosol formation and evolution in the stratosphere. II. Sensitivity studies and comparison with observations. *J. Atmos. Sci.*, vol. 36, No. 4, p. 718–736.

95. Wang Wei-Chyung, G.A. Domoto. 1974. The radiation effect of aerosols in the earth's atmosphere. *J. Appl. Meteorol.*, vol. 13, No. 5, p. 521–534.

96. Weare, B.C., R.L. Temkin and F.M. Snell. 1974. Aerosol and climate:

some further considerations. *Science*, vol. 186, No. 4166, p. 827–828.

97. Welch, R.M. and W.G. Zhunkowski. 1976. A radiation model of the polluted atmospheric boundary layer. *Preprint Dept. Meteorol.*, Univ. Utah, Salt Lake City. 42 p.

98. Yamamoto, G. and M. Tanaka. 1972. Increase of global albedo due to air pollution. *J. Atmos. Sci.*, vol. 29, No. 8, p. 1405–1412.

99. Zhunkowski, W.G. and R.L. Weichel. 1971. Radiative energy transfer in haze atmospheres. *Beitr. Phys. Atmos.*, vol. 44, No. 1, p. 53–68.

100. Zhunkowski, W.G. and K.N. Liou. 1976. Humidity effects on the radiative properties of a hazy atmosphere in the visible spectrum. *Tellus*, vol. 28, No. 1, p. 31–36.

101. Zenchelsky, S. and M. Youssefi. 1979. Natural organic atmospheric aerosols of terrestrial origin. *Rev. Geophys. and Space Phys.*, vol. 17, No. 3, p. 459.

Index*

*Numbers appearing in the Index refer to those in the original book. These are mentioned in the left-hand columns in the present text.

280

Printed in India